BIG DAMS : CLAIMS, COUNTER CLAIMS

BIG DAMS : CLAIMS, COUNTER CLAIMS

B.D. DHAWAN

1990
Commonwealth Publishers
New Delhi

BIG DAMS : CLAIMS, COUNTER CLAIMS

Edition 2024

ISBN 81-71769-063-7

Published by :

COMMONWEALTH PUBLISHERS (P) LTD.

4831/24, Govind Lane,
Ansari Road, Darya Ganj,
New Delhi - 110002
Ph. : 23272541, 23242541
e-mail : commonwealthpublishers2016@gmail.com

Distributed by :

ARJUN PUBLISHING HOUSE

4831/24, Govind Lane,
Ansari Road, Darya Ganj,
New Delhi - 110002
Ph. : 23272541, 23242541
e-mail : arjunbooks2001@gmail.com

Laser Typesetting by :
Dimple Computers
Delhi

Printed at :
Milan Enterprises
Delhi

IN AFFECTION AND GRATITUDE TO

My Maternal Uncles, Aunts and My Dear Sisters

Preface

This book is a collection of some noteworthy writings on the current big dam debate. The papers presented in this collection would give the curious reader some idea of the numerous aspects of this debate. Many of these papers are already published. Bringing them together in one place is an attempt to focus attention on the diverse issues underlying the debate. It is hoped that this would generate wider and more informed participation in the controversy, and hopefully to its eventual resolution.

The papers are all of recent period, beginning with my paper 'Mounting Antagonism towards Big Dams', published in May last year in the columns of *Economic and Political Weekly*. A discussion ensued immediately in this journal. Besides Dr. Arun Ghosh taking up the Narmada controversy in his own column 'From the Ivory Tower', two more papers appeared in the Review of Agriculture section of its September 1989 issue, one by Ramaswamy R. Iyer ('Large Dams : The Right Perspective') and the other by this writer ('Major and Minor Irrigation Works : Cost Aspects of the Controversy'). This writer then decided to collect these *EPW* papers, along with the discussion papers, into a volume for the benefit of wider readership. While doing so, he decided to include other papers/writings which had appeared contemporaneously in other journals.

Simultaneously, I approached a select group of very knowledgeable persons to contribute to the volume their own recent writings on the subject. The response was not encouraging. As a result, some kind of imbalance has remained in the present collection. Some well-informed critics of the big dam idea do not sufficiently figure in it. In a sense, however, this over-representation of the votaries of the big dam is not to be lamented. Their standpoint has remained grossly under-represented in the media. Rather, they are late entrants to the debate to defend their position.

Including the introduction, this volume contains twelve chapters. It is well nigh impossible to categorise the papers. Some question the big dam idea in its entirety, some support the same to the hilt. There are others which provide a look at both sides of the coin, although one may discern streaks of their authors' predilection one way or the other.

Among the numerous acknowledgements, the foremost thanks must go to the contributors. I am also grateful to Mr. Krishna Raj, Editor of *EPW*, for cooperation and encouragement in the production of this volume. As usual, my colleagues, notably Dr. Suren Navlakha and Dr. Kanchan Chopra, evinced keen interest in the whole endeavour. My friend, Shri B.D. Kapoor, has been a source of support in bringing out this volume. Finally, I am indebted to the Institute of Economic Growth, whose environs and infrastructure were of great help in the pursuit of this venture.

March 1990 B.D. DHAWAN

Contents

CONTRIBUTORS

KANCHAN CHOPRA (b.1944) is Professor of Economics at the Institute of Economic Growth, Delhi. Her initial research was in the area of project evaluation and investment planning, in the particular context of the irrigation and forestry sectors. Of late, issues related to environmental degradation and participatory development processes have occupied her attention. She has published extensively in academic journals. She is the author of the following books : *Agricultural Development in Punjab : Issues in Resource Use and Sustainability* (1990), *Participatory Development : People and Common Property Resources* (with Gopal K. Kadekodi and M.N. Murty), *and Dualism and Investment Patterns : An Analysis of Regional Contrasts* (1972).

SHRIPAD DHARMADHIKARY is an activist researcher, working earnestly for the last six years in the submergence zone of the Sardar Sarovar Project in the Madhya Pradesh part of the Narmada basin. His professional background is that of an engineer (mechanical), with a B. Tech degree from IIT, Bombay. He has deep interest in development issues, especially those of energy and water resources. He has explored, through computer simulations, future paths of sustainable development. He is associated with two voluntary organisations, namely, NARMADA DHARANGRAST SAMITI (Maharashtra) and NARMADA GHATI NAVNIRMAN SAMITI (Badwani, M.P. -- 451, 551) -- his association with the latter is fulltime and he can be reached via the address of that organisation.

B. D. DHAWAN (b. 1935) is Professor at the Institute of Economic Growth, Delhi, with which his professional research career as an applied economist began in 1961. His earlier research work was in agro-industries and satellite communications. Since the mid-seventies, he is concentrating on irrigation. Besides several journal articles, he is the author of the following books on irrigation : *Studies in Minor Irrigation* (1990), *Studies in Irrigation and Water Management (1989), Irrigation in India's Agricultural Development* (1987), *Economics of Groundwater Irrigation in Hard Rock Regions* (1986), and *Development of Tubewell Irrigation in India* (1981).

RUSI ENGINEER (age 45 years) is a versatile person, with degrees in Accounting and Auditing, Philosophy and Economics (from London School of Economics and Political Science) and Psychology – also a diploma in Film Production (New York University, 1972). Earlier, he was Senior Lecturer in National College of Bombay University (1974-78), and the Western Region Organiser for World Wildlife Fund during 1978-81. Much of the eighties he was with *Business India* magazine, first as its Deputy Editor (1982-83) and later as Editor (1984-89). He has written a number of articles on diverse topics, and two booklets viz. *Soil : Our Daily Bread* and *Sardar Sarovar Project : An Economic, Environmental and Human Disaster* -- also an audio-visual production titled 'Our Vanishing Forests'. He plans to start a high-quality new journal *Ecology, Environment, Sustainable Development.*

ARUN GHOSH (b. 1923) is an economist Member of the newly constituted Planning Commission, New Delhi. Earlier, he taught Economics at the University of Lucknow, and has been associated with the framing of the first official estimates of India's national income and the economic classification of the Central Government budget. He has been the Alternative Executive Director for India on the World Bank and the IMF, and a Secretary to the Government of India and Vice-Chairman of the West Bengal State Planning Board. He is the author of *Prices and Economic Fluctuations in India 1861 to 1947,* and *West Bengal Landscapes.*

RAMASWAMY R. IYER (b. 1929) is now a Member of the Public Enterprises Selection Board, New Delhi. Since he joined the Indian Audit & Accounts Service in 1953, he has held very many senior positions in a number of Economic Ministries in the Government of India. He was Secretary of the Economic Administration Reforms Commission (1981-1984). He was Senior Adviser in the International Centre for Public Enterprises (Yugoslavia) for one year (1984-85).

He retired in 1987 as Secretary, Ministry of Water Resources. During the period of about two years as the Secretary, he attempted a reorientation of the Ministry from a preoccupation with major irrigation projects to a concern with issues of water resource development and water management. Soon after joining the Ministry, he initiated the preparation of a National Water Policy document and was largely responsible for drafting it.

JASVEEN JAIRATH (b. 1950) is scientist fellow at the National Institute of Science and Technology and Development Studies, New Delhi. She has a multidisciplinary background : electrical engineering degree from the Birla Institute of Technology and Science, Pilani ; Masters degree in Economics from New School for Social Research, New York, and Ph.D. in Economics from Jawaharlal Nehru University, New Delhi. Earlier she has served as assistant design engineer (1973-77) and economist (1977-79) in the Irrigation Department of Punjab State. In 1985, she collaborated with Dr. Krishna Bhardwaj in a brief ICSSR survey *Studies of Irrigation in India – Alternative Perspectives*. Since 1985, she is publishing her research findings in *Economic and Political Weekly*. At NISTADS she is researching on two projects : (1) Technical Change in Machine Tool Industry in India, and (2) Efficiency, Productivity and Equity of Resource Use -- A Case Study of Irrigation Systems in Haryana.

ASHOK K. MITRA (b. 1940) is Reader at Gokhale Institute of Politics and Economics, Pune, with which his association dates back to 1972 when he obtained his doctoral degree in Economics from Poona University. At the Institute he has researched on a variety of problems, including rural electrification, agricultural employment, rural development and irrigation. He has by now developed very strong research interest in issues of water resource development and management. He is undoubtedly a well-published scholar in *Artha Vijnana* (a quarterly journal of the Gokhale Institute) and the *Indian Journal of Agricultural Economics*. He has authored *Economic Benefits of Rural Electrification in Maharashtra —A Cost-Benefit Analysis of Rural Electrification* (1980) and *Underutilisation of Irrigation Potential in the Context of Protective and Productive Irrigation* (forthcoming).

CHANDRA SHEKHAR PANT (b. 1970) is a bright undergraduate student of engineering at the College of Technology, Pantnagar (U.P.). He represents the young student community that has been roused by the environmental movement in the country, and which is quite willing to give a helping hand to activist organisations with their time and energies.

NIRMAL SENGUPTA (b. 1946) is Professor at the Madras Institute of Development Studies, Madras, for which he is also the Acting Director presently. He is a product of the Indian Statistical Institute (ISI), from which he obtained his Masters of Statistics (1968), Diploma in Econometrics (1969) and doctoral degree (1987). After short stints in ISI, Calcutta, and Bihar State Planning Board, he had long association with A.N. Sinha Institute of Social Studies, Patna (1970-83). His research interests are wide, covering ethnic minorities, industry, agriculture and irrigation. He has several contributions to his credit. He is author of *Our Country : An Economic Introduction (1979, in Bengali), Destitutes and Development : A Study of Bauri Community in Bokaro Steel Region, Fourth World Dynamics : Jharkhand* (ed., 1982), *and Indigenous Property Relations and Development Strategies: A Study of Water Management in India and Philippines* (in press).

R.B. SHAH (b. 1932) has recently retired as Chairman, Central Water Commission (CWC), with which his association began 33 years earlier. He has several technical papers to his credit. His tenure with CWC enabled him to participate in all aspects of water resources development in India. As Chief Engineer, Planning and Investigations, he investigated and prepared project reports for a number of water resources projects on irrigation, hydro power, and water supply. As a Member of the CWC, he was incharge of techno-economic appraisal of all the major, medium and multipurpose projects. As convenor of the Planning Commission Working Group on Major and Medium Irrigation for the Eighth Plan, he prepared the report of this Group. Besides visits abroad and interface with external agencies like World Bank, he has field experience in the construction of micro hydel projects in Bhutan, and overseeing the construction of Rajghat and Bansagar multipurpose projects.

MANU SHROFF was till recently Editor of the prestigeous economic daily *Economic Times*. He was educated at the University of Bombay and the famous London School of Economics. He has held professorship at the Indian Institute of Management, Ahmedabad, before joining the central civil service. He has been inter alia Economic Adviser in the Ministry of Finance, Adviser to Planning Commission, Additional Secretary (Banking), and India's Alternate Executive Director on the World Bank.

A. SUNDAR is a mature engineer with very keen professional interest in the development and management of our water resources for irrigation purpose. Earlier, he was Professor at the Indian Institute of Management, Bangalore, from where he edited the quarterly *WAMANA*, a noted journal on water resources. He is presently Chairman and Executive Director, Irrigation Research Academy, Bangalore.

K. C. Thomas (b. 1922) is a retired Chairman of the Central Water Commission. He obtained his M.Sc. and D.Sc in engineering from the famous Carnegie Institute of Technology, USA. He joined CWINC (now CWC) in 1949, rising from Assistant Engineer level to its Chairman (1978-80). In between, he has been both Reader and Senior Professor at the Water Resources Development Training Centre, Roorkee (5 years); Executive Engineer, Hirakud Dam Project (5 years); Member and Chairman, Himachal Pradesh Electricity Board (4 years); and General Manager, Chukha Hydro Electric Project (3 years). Since retirement, he has done consultancy service in India as well as abroad, besides being Chief Technical Adviser to the United Nations (1983-87).

1

Introduction

B D Dhawan

I

The big dam issue has indeed become very controversial. That should not be too disconcerting a thing, though. Controversies on issues of public importance are the life-line of democratic polity. Therefore, persons at the helm of affairs of a nation/state should not only view such controversies as legitimate and healthy, but be favourably disposed to these. They need to be particularly responsive to public criticism when it is directed against the official line of thinking. Rather than trying to prevent these or wishing them away, they can, through appropriate information and right posers, make these more purposive, and less intense in bitterness, less acrimonious, less unpleasant and of course less durable.

In this regard, the imperative of an 'open government' policy hardly needs to be emphasised. Such openness, to begin with, thwarts the hustlers of pet projects who have little regard for objective planning processes. Secondly, it obviates the need for recourse to muffling dissensions which only forces the dissenters to resort to extreme steps like 'fast unto death', 'rasta roko', mass rallies, etc. Thirdly and most importantly, once the relevant information germane to a controversy is freely made available to the seekers of such information at least from the world of media and academia, an ethos of informed discussion instead of polemical debate is nurtured. Discussion brings the central issues in bold relief, while relegating non-issues to the background. In ultimate analysis, such interventions from media/academia hasten the process of resolution of a controversy. This book is a small endeavour in this direction, for which the prospects are now propitious following the recent changes of regimes at the Centre and in many states.

Many misgivings about major irrigation works have prominently figured in our print media during the last two years. Their rallying point has been the highly controversial Tehri dam on the Bhagirath (Ganga basin) and the gigantic Sardar Sarovar and Narmada Sagar reservoirs on the main Narmada river of Central India. There is no denying that approval for the construction of these projects was not only given in great haste but the very process of clearing them was not above-board. No wonder, this hustling has made the media by and large hostile to both these projects and their planners. In the absence of any serious attempt by the planners thereafter to respond to their critics in the public and planning fora, the general climate has become very susceptible to all kinds of apprehensions and misgivings about big dams expressed anywhere and by anyone. When hustlers rule the roost, they adopt dismissive postures rather than engage in an open debate with their trenchant critics.

Fears of ecological disaster and deep concern for the dam oustees have led to the formation of many activist groups against the big dam idea lately. They have been trying not only to win wider public support for their cherished cause but in the face of dismissive attitude of the authorities are also engaged actively in thwarting the ongoing construction work on the above mentioned dams.

The task of resolving the controversy has now become rather stupendous. The ranks of anti-big dam forces have swelled enormously, while huge sums of money have already been spent on the controversial projects. Were it indeed to be concluded that these projects are of dubious worth, a question naturally follows : what might be the salvage value of these investments ? Should these be treated as the price to be paid for restoring Indian planning process on right rails ? Evidently, this would apply not only to irrigation but to all mega projects that got initiated without due regard for planning processes they deserve.

II

Before introducing the reader to the papers in this collection, it would be pertinent to mention that big dam controversy is a vastly complex and many sided issue. Its resolution demands a meeting of many minds, many disciplines and many concerns. What are our compulsions for developing irrigation ? Cannot the new dryland farming technology meet our rising requirements of food ? If irrigation has still an expanding role to play, what options do we have for its development? Instead of big dams, can we not tap our surface water resources through alternatives like small-scale reservoirs, farm ponds, micro dams, etc. ? How do these options compare in cost-effectiveness ? Have we exercised our options wisely so far ? In what direction may we boldly move in the years ahead, including the Eighth Five Year Plan under preparation for the period 1990-95 ?

Admittedly, no definitive and generally acceptable answers exist to all these and allied questions, although these are so germane to the crystallisation of issues involved in the discussion about big dams[1]. In what immediately follows therefore a modest attempt is made to grapple with these questions. Hopefully, this will facilitate, not hinder, a critical appreciation of the diverse viewpoints presented in this collection.

Compulsions of Irrigation Development

Protecting rural population from the ravages of periodic droughts is no longer a dominant consideration for developing irrigation in India. Rapid increase of production is now an overriding motivating factor behind irrigation development. Indeed, the production aspect has become the most pressing objective for the planners, even at the cost of relegating the issues of equity, long term security of the basic resources and the eventual sustainability of the programmes of production into background, or so it would seem. It is precisely these neglected issues which are brought into the forefront by the increasing concern for environmental preservation, ecological balance and social justice, in the process offering a good deal of resistance to the pure pursuit of production objective.

Unlike the conditions prevailing in high income societies, foodgrains, edible oils and sugar (including palm and cane jaggery) absorb a lion's share of household income in India. Therefore, as the

1. My main familiarity is with groundwater economics literature to which I contributed a bit over the last two decades. Whatever irrigation research I have done on surface water sources is far from the holistic tradition. My own thinking in this area is still in an evolving phase.

Indian population grows relentlessly at an annual rate of a little over 2 per cent, one has to provide for a commensurate additional production of food articles. To further take care of the income effects generated by the planning effort, this provision has to be kept above the rate of expansion of population, say, of the order of 2.5 per cent for cereals and 3 to 4 per cent for edible oils, sugar and pulses for which the income elasticity of domestic demand, unlike in the advanced countries, is still very sizable. Evidently, failure to keep pace with our growing basic needs can be distressing in the short run and catastrophic in the long run.

For a nation of continental dimensions, any substantial reliance on food imports too is to be ruled out, however cost-effective such an option may seem in comparison to self reliance. This import route is surreptiously commended by nations who have surpluses of food to dispose of in the world market. But when it comes to production of essentials of life/economy/defence, the notion of comparative advantage across nations is unsound and politically unsuited to a large but poor country like India. For us, this notion is much more relevant for planning optimal crop production across the states within the Indian Union.

Precisely while making such domestic investment decisions, due preference has to be accorded to options that result in higher, less uncertain and sustainable gains in food production. Can dryland technology alone fulfil this vital role ? I am raising this question because the proponents of this technology, especially some ICRISAT scientists, have been overprojecting it while making a critique of irrigated agriculture. Let me concretise an element of even deliberateness in this overprojection. While berating Indian irrigation for not ushering in the promised era of high crop yields (4 to 6 tons per crop hectare), these champions of new dryland technology fail to point out that in actual field conditions this technology has raised yields from a level of 4 to 6 quintals/ha to 8 to 12 quintals/ha only, and not to 25 quintals/ha which is the claimed potential of the technology. (One ICRISAT scientist, carried away by his own enthusiasm, even projected this potential to 40 quintals/ha, if only in his oral presentation at a recently held high-level national seminar).

The role of dryland farming technology needs to be understood correctly. Its promotion has an unquestioned case, in the sense that our water resources cannot suffice for practicing wholly irrigated farming on every hectare of our arable land. Accepting for the moment the expert assessment that our water resources ultimately may suffice for providing irrigation cover to only 55-60 per cent of the crop area, we do need a promising dryland technology for the rest of the crop area, so as to both realise high output from such crop area and provide benefits to the farmers there. Therefore, adequate investment in dryland research and its propagation has to be made in our planning. From all accounts, we have so far underprovided for this option and must correct it by suitably stepping up the outlays in the coming years. Irrigation planners should not resist such a step-up. After all, some of these outlays, say on watershed management, do have positive external effects on irrigation activity itself – by improving groundwater availability and arresting the siltation process of the reservoirs.

But promotion of dryland farming cannot be by the exclusion of irrigation development, as was indeed recommended by the Indo-US Sub-Commission on Agriculture in 1987. Whenever Indian agriculture has witnessed a short spell of stagnation, doubts have been expressed about the overall planning effort for this sector, especially about the role of irrigation vis-a-vis dryland technology. Writing in the wake of one such a recent spell of stagnation (1983-86), this Sub-Commission endorsed the *Combined Report of the US Dryland Farming Team and the Economics Team Visits to India*. The Economics Team warned of a 'coming quiet crisis' unless dryland sector was enabled to contribute an additional 49 million tons in foodgrains by 2000 A.D. In a veiled manner, it urged the World Bank and the USAID to shift the gear from irrigation development to dryland development in India. This author has shown elsewhere (Dhawan 1988) what was amiss with the assumptions as well as the perceptions of the writers of this Report. Here it would suffice to emphasise the following three empirical facts :

(i) There is no evidence of any firm uptrend in overall crop yield for dry lands in India.
(ii) There is a noteworthy uptrend in overall crop yield for irrigated lands in toto as well as for lands watered by canals fed by river flows.

(iii) Instability in rainfed yields is much more than in irrigated yields.

In a country of vast diversities it is not prudent to propound (a) a uniform policy for the entire country and (b) some particular alternative to the exclusion of all others. As a matter of fact, both dry and irrigation technologies have their due roles to play, with change of emphasis from region to region. So, it would be suicidal to depart from the policy of irrigation as a major feature of crop planning, given the existing set of requirements and possibilities. In this connection, we might take note of an emerging realisation among some truly perceptive promoters of dry farming technology, whereby they are veering round to the view of "life saving" irrigation. Researches have shown that even a single, light irrigation can appreciably boost land productivity. Thus experiments on a staple grain of the Deccan (i.e., jowar) indicate a step-up in yield from 11 to 19 quintals/ha with a "life saving" irrigation of just two inches (5 cm).

This needs to be noted especially by our agronomists and plant breeders who tend to run down surface irrigation in general and public canals in particular when these are used to provide an extensive protective cover rather than to facilitate full realisation of the yield potential of new crop varieties. They fail to appreciate that limited water resources, when distributed extensively, do maximise total production from land as well as water resources, besides providing spatial equity in gains from public investments. Nor have they reckoned with the dynamics of protective irrigation. As the benefits of irrigation accrue, farm incomes and savings rise. As a good deal of the accumulated savings are ploughed back in agriculture, more so in the form of investment in private well irrigation, land productivity rises further. The access to underground water improves the quality as well as quantity of irrigation to the canal farmers, as also restores the groundwater balance disturbed by the seeped-in canal waters.

Options in Water Utilisation

As already alluded, our irrigation resource is far more limited than the arable land resource. Further, any water resource, whether this be on land surface or underneath it, becomes available for irrigation only after some expenditure is incurred in capital as well as current cost terms. The composition of capital and recurring costs varies radically between surface and groundwater based irrigation works. Whatever be this composition, the pertinent reality to remember is that irrigation as such is a costly proposition. Therefore, choices are between costly options, and not between "low-cost" groundwater irrigation and "high-cost" big dam irrigation. The illusion of low-cost options arises when costs are not judged in a comprehensive framework.

Furthermore, clear-cut choices are not always easy to make owing to hydrological linkages between water bodies. That is to say, development of surface irrigation works like canals and tanks disturbs groundwater balance as the seeped-in canal/tank water becomes an additional source of groundwater recharge, over and above the natural groundwater recharge from rainfall and rivers. Likewise, massive groundwater extraction can adversely affect surface water availability. Some hydrologists indeed view ground and surface waters as indivisible, and while one may not wholly agree with this, it poignantly underscores the inter-connectedness of these two sources. This inter-connectedness, alongwith the tag of time fastened to water availability, renders irrigation/water resource a complex resource. The complexity is compounded all the more by water's perplexing interactions with soil and plants. For instance, a plant may wilt because of soil tension even though considerable quantity of water may still be there in its root zone. Likewise, afforestation may improve groundwater recharge by curtailing surface run off, but with a marked decrease in *net* availability of both groundwater (due to groundwater use by trees) and in surface flow to a downstream reservoir. Among the real gains from such afforestation of denuded hill slopes is of course the reduction in inflow of silt and debris into the reservoir/river.

In view of the externalities between ground and surface waters their conjunctive use for irrigation has become an axiom for the water planners everywhere. The Indian case for such conjunctive development is further strengthened by the fact that their combined irrigation potential is less than our irrigation requirements. This case is not altered even if one were to accept the questionable claim now being advanced by the Central Groundwater Board, that the ultimate irrigation potential from our groundwater resources is twice the earlier estimate of 40 million hectare area. The veracity of this claim has been questioned by the author elsewhere (Dhawan 1989, 1990a, 1990b).

Thus the issue meriting consideration for water planning in India is not of mutually exclusive choices between dryland farming and irrigation, or between surface and groundwater irrigation. It is really one of small versus large irrigation works for harnessing our surface water flows. The official expert assessment is that, for a variety of techno-economic considerations, a major part of these flows is bound to remain unexploitable for irrigation purpose. One can question this assessment and argue that a good fraction of this unexploitable water resource need not flow to the Indian ocean. Instead, some of it may be harnessed through the following options :

(1) constructing farm ponds;
(2) intensifying soil and water conservation works;
(3) providing better tree cover to the hillsides and hill tops;
(4) transferring surplus flows from one river basin to another.

As per the above expert assessment the irrigation potential from our surface water flows is of the order of 73 million ha of crop area, of which 15 million ha is placed against small scale or minor schemes (each with a culturable command area of less than 2000 ha) and the rest from major and medium projects. The inter-basin transfers, which can only be through major projects, are potentially capable of enhancing the irrigation potential to almost 108 million ha, as per the preliminary investigations. In addition to this, farm ponds and water conservation works may also enhance minor irrigation potential from both surface water and groundwater sources.

While soil conservation works enhance the sustainability of the downstream irrigation works, the afforestation of the upper high rainfall catchments may modify the 'hydrograph' of river (i.e., the time distribution of river flows in a year). That is to say, the peakedness of flows during the main monsoon months may be attenuated, simultaneously enhancing the flows in dry period season in quantum as well as duration. This gain in temporal distribution, even at the expense of overall reduction in the mean flow of a river, is highly welcome, since it brings the temporal demand for water (for irrigation as well for other purposes) in alignment with natural water availability. This gain is in addition to the normal gains from afforestation programme, as mentioned above.

That an integrated watershed management programme is an extremely worthwhile investment option should be beyond a pale of doubt. However, two reservations in this regard need mentioning. First, the tall claim of some soil and water conservationists that this instrument can flatten the hydrograph of a main river to a point where floods in rivers would become a thing of the past is grossly excessive. The attenuation in peak flow is governed much less by the extent and nature of soil cover than by the soil depth of the hillside in which rain drops are to be held or stored up. For instance, 10 to 12 inch of dry soil depth is required to store water from a rainfall spell of two inches, but when 10 to 12 inches of rain occur in a row over three to four days, it would require a soil depth of 4 to 5 feet to impound the same in the hillside soil. Deep soils are found in the valleys (where the eroded soils accumulate), and not on hillsides to be brought under forest cover.

Second, the gains from watershed management programme become a pale shadow of the potential gains whenever the programme is poorly planned and implemented, which, alas, has been our wont. People's participation in this programme is expected to improve the implementation. This is a relatively new idea and it has shown good results at a few places it has been tried. Time alone will show how this promising route would work in practice when it is extended in a big way to large tracts.

Progress in Major Irrigation

From the viewpoint of expansion attained in "created" irrigation potential, the progress is impressive in both major (including medium) and minor irrigation segments of the Indian irrigation sector. At the commencement of the planning era in 1951, the capacity of major irrigation segment was rated to be of the order of 9.7 million ha which has now been reassessed at 8.6 million ha. This derating has been done very recently by the Central Water Commission, alongwith a downward revision of the achievements of the Sixth Plan. As a result, the total irrigation potential stands revised from 30.01 million ha to 27.64 million by the end of the Sixth Plan. Owing to this downward adjustment, the new capacity of 3.13 million ha anticipated from the Seventh Plan has hardly meant much addition to total capacity. Yet, the progress during the four decades of planning is of historic importance, in the sense that additions to capacity in this period were at least three times those that occurred during the long British rule (varying from 10 to 20 decades in different parts of India).

While no long-run signs of deceleration in progress are visible, there are also no signs of accelerated development in major irrigation. The progress has varied from Plan to Plan, the best being in the Fifth Plan period (1974-75 to 1977-78) when nearly 4 million ha capacity was established in four years time. Plan-wise additions are as follows :

1. First Plan	2.48 million ha
2. Second Plan	2.14
3. Third Plan	2.24
4. Three Annual Plans (66-69)	1.53
5. Fourth Plan	2.61
6. Fifth Plan (4 years duration)	4.01
7. Two Annual Plans (78-80)	1.80
8. Sixth Plan	2.11 (reassessed)
9. Seventh Plan	3.13 (anticipated)
10. Eighth Plan	6.5 (target)*

* As recommended by the Working Group on Major and Medium Irrigation.

Major irrigation works have been much berated for their deficiencies, more so about unutilised capacity declared "created" by the planners. It is not my intention to catalogue here the various criticisms of major irrigation works in general and of big reservoir works in particular. My own thinking in this regard, including a critique of the critics' standpoints, is contained in my write-ups in this collection. I, therefore, propose to concentrate here on the intractable problem of project spillovers both in time and costs. This is a general malaise of most public sector projects. But before that, some general observations would be in order.

It requires extraordinary skills to harness mighty rivers for mankind's benefits. Those who acquire these skills through talent and training may genuinely come to believe that major irrigation works which they help to create are nothing less than a boon to the society. But not even God's blessings come unmixed to man; there are pluses and minuses too. As it turns out, the minuses of major irrigation works have been far more vigorously commented upon than pluses.

While senior irrigation engineers have been carrying on the planning of major irrigation works on the premise that the pluses far outweigh the minuses, the myraid critics of such works have questioned this premise. Some rank bad projects, like the Tawa and the Sarda Sahayak, have been so poignantly utilised by some critics as to tarnish the entire image of major irrigation segment. Since engineers have neither tried to check the veracity of their own premise nor shown willingness to engage in a meaningful dialogue with their critics, the prevalent widespread impression is one of high, soaring costs and low, declining benefits from major irrigation works. Therefore, few take seriously the rosy benefit-cost ratios projected for the new projects submitted to the planners.

The Project Spillover

The long gestation period has been held as an important drawback of investments in major irrigation works (vis-a-vis minor works) ever since the commencement of planning in the country. A major irrigation project, even when implemented under good conditions of preparatory investigative work and reasonable project funding thereafter, can take 10 to 12 years of time for completion. Therefore, a carryover problem, from one Five Year Plan to another, is unavoidable in such projects. Also, because of the compulsion of maintaining a steady rate of expansion in major irrigation capacity, one cannot wait for the prior completion of all the ongoing projects before starting new works. In other words, some ongoing projects, alongwith some newly started projects, would exist in every Five Year Plan (except the First one) until further scope for setting up new projects is more or less exhausted.

Therefore, it is naive to urge planners that in launching a Five Year Plan no provision for new major irrigation projects should be made *till all the ongoing ones are completed*. However, the compulsion of spillover should not become a licence for starting any number of new projects. The planners have to maintain a right balance. The imperative of such balancing becomes all the greater when projects are getting unreasonably delayed and serious time overruns are occurring.

Between the First and the Seventh Plan 264 major irrigation schemes (besides 1088 medium irrigation schemes) were taken up for execution. The spurt in starting new projects came during the Fifth Plan – this is the Plan known for the best achievement in major irrigation development – when 73 major and 331 medium new works were started. Of the nearly 140 ongoing major projects which are likely to remain incomplete by the end of the Seventh Plan period (March 1990), as many as 55 are of the Fifth Plan vintage. But no less than 32 are of pre-Fifth Plan days. These two categories of incomplete projects, whose commencements are indicated below, speak volumes for the time overruns in Indian water planning for the major irrigation segment.

Spillover from :	No. of major ongoing projects
First Plan	2
Second Plan	2
Third Plan	11
Three Annual Plans (66-69)	3
Fourth Plan	14
Fifth Plan	55

Despite serious problems in executing projects underway, many new projects were allowed : 15 in the Annual Plans of 1978-79 and 1979-80, 41 in the Sixth Plan, and 18 in the Seventh Plan. When the states show greater interest in starting new projects than completing the ongoing ones (and in maintaining the already completed projects), the limited investible funds earmarked for irrigation sector get thinly spread on too many projects. Large chunks of the inadequate allocations released for a project get absorbed in defraying establishment costs (that are inflexible on the downward side but very flexible upwards following the rise in dearness allowance (DA) claims owing to price inflation), while the real works component of a project in a year is implemented far less than the desired level determined under the project schedule. It is in this sense that one might take the oft-repeated allegation that Indian irrigation bureaucracy has become a great beneficiary of the investment being made in major irrigation projects.

Of the pernicious consequences of such delayed completion of irrigation works, two may be mentioned here. First, it leads to distortions in the whole planning process. Too many ongoing mega projects, each implemented at a snail's pace, generates a tremendous pressure for resource diversion from other activities, e.g., dryland farming and medium and minor irrigation works of the public sector. In other words, a *crowding out* effect is generated. Second, the true cost of establishing irrigation service shoots up. I am not referring to the mind boggling cost overruns of projects. Stated in nominal terms, without any due adjustment for the persistent price inflation, they rather exaggerate the project cost escalation. Due deflation of nominal costs would yield a truer resources cost of a project. But the lengthening of the gestation period implies high opportunity cost of production/income benefits deferred by many years for a poor economy.

It must be clarified that spurts in major irrigation development are not closely related to the experience of unprecedent droughts at that time. A crisis of drought leads to crash programmes of quick-maturing minor irrigation works in general and of tubewells in particular. Of course, major irrigation works nearing completion at the time of a drought do also get priority in plan allocation. It is one thing to assert that major irrigation works play a useful role in mitigating the potential ravages wrought by drought. But this does not trigger off new major irrigation schemes of long gestation period. It is the politics of food which needs to be comprehended to fully grasp the distortions that have crept in major irrigation planning in India.

The two droughts of the mid-sixties did much to the development of groundwater irrigation. But irrigation came to acquire a priority of sorts in Indian planning mostly in the wake of India's action in Bangladesh. Then it received political backing that came from the realisation of new compulsions and interests. Following the grave Indo-American differences arising out of the Indo-Pak conflict on Bangladesh issue, the political compulsions for achieving self-reliance in foodgrains dawned afresh on the Indian leadership keen to play some part in the global and regional geo-politics. Knowing well that the major grain exporter of the world uses its huge grain surpluses as an instrument of foreign policy, wisdom demanded not to expect a repeat performance of magnanimity in food aid of the kind witnessed during the mid-sixties.

Close relation exists too between increasing emphasis on irrigation development and internal politics in two principal ways. First, irrigation is a powerful and speedy tool for enhancing as well as protecting farm incomes in low and medium rainfall regions of the country. This compels state politicians in these regions to respond to farmers' aspirations for irrigation development. Electoral politics and appeal to populism generate excessive zeal for starting several irrigation schemes (some not even approved by the Planning Commission) well beyond the available investment funds. Secondly, ambitious politicians in the States have realised the importance of surplus foodgrains. If they belong to states with surplus grains, they can exercise clout in the domestic political arena in comparison to those in deficit states who feel hampered.

Excessive zeal for irrigation as such has led to faulty irrigation planning, as witnessed especially during the Fifth Five Year Plan which began close on the heels of Indo-Pak war of 1971 when too many irrigation schemes were taken up. How the then-incumbents of the Yojana Bhavan allowed this serious fault in irrigation planning to overtake them needs a fuller probe. When an idea is placed in the centre of things and thereafter continues to absorb a large chunk of investible resources on the plea of priority for ongoing projects, it causes heart burning among those who feel frustrated in promoting their own scheme of things for improving the economy and society.

Proposals for Eighth Plan

A ticklish issue facing the new incumbents in the Yojana Bhavan is that of shelving some of the ongoing projects of dubious worth. The number of ongoing projects likely to remain incomplete by the beginning of the Eighth Plan (1990-95) is exclusive of the following 17 projects recommended for deletion by the Working Group on Major and Minor Irrigation for Eighth Plan :

1. Four *unapproved* projects of the Cauvery basin of Karnataka (Kabini, Hemavaty, Harangi and Varuna Canal)

2. Three projects in Gujarat (Heran, Zankhari and Sidhumber)
3. Six projects of Maharashtra (Lower Wardha, Dudhna, Nira Deoghar, Ghosi Khurd, Lendi, Lower Penganga)
4. One project each from Andhra Pradesh (Polavaram), Haryana (Ladwa), Kerala (Beypore Puzha) and U.P.(Kishau dam).

While the Heran project (Gujarat) is scratched foreover, in others expenditure upto 3 per cent of the project investment has been already incurred. The three following ongoing projects have been proposed for transfer to other plan heads :

1. Thein dam in Punjab, to power sector
2. Tehri dam in U.P., to power sector
3. Additional spillway of Hirakud (Orissa), to "Restoration and Improvement".

The ongoing irrigation projects, numbering 142, have about 13 million ha of *unrealised* potential. Though 45 new project starts are recommended by the Working Group, bulk of the targetted 6.5 million ha of additional capacity envisaged by the Group would come from ongoing projects, of which 95 are anticipated to be completed by the end of the Eighth Plan *provided* some Rs. 14,700 crores are exclusively earmarked for them. The target is certainly ambitious in comparison to achievements during the recent plans. If this target is brought down to the neighbourhood of 4 million ha, the case for starting 45 new irrigation projects gets weakened. For, the 13 million ha capacity from the ongoing projects would suffice our requirements in the subsequent two Plans also. In other words, we can afford to defer the construction start on these 45 new projects to Ninth Five Year Plan.

III

About the Papers

'Mounting Antagonism towards Big Dams' by this writer constituting Chapter 2 in this volume, was published in the 'Perspectives' column of *EPW* of May 20, 1989. Its key idea of lowering dam heights had been first presented by him in the 'Viewpoint' column of the *Economic Times* of December 5, 1988, under the title 'The Way Out of the Narmada Imbroglio'.

Essentially, the paper is an attempt to do a stock-taking of the increasing questioning of the assumptions underlying the irrigation planning in general and of major irrigation works backed by big reservoirs in particular by environmentalists, soil conservationists, and social and natural scientists. It laments the water planners' disinclination to engage in an open dialogue and discussion with their critics. It suggests a via media solution to all highly controversial proposed big dams, that is, scaling down their height so as to accommodate substantially the genuine apprehensions of the critics in respect of (1) loss of forest cover, (2) dam-induced seismicity and (3) rehabilitation of large number of dam oustees.

The aforesaid paper came under scrutiny and attack in the 'Discussions' column of *EPW*. Three such discussion papers are included in chapters 3, in the order in which these appeared in *EPW*. Ashok K. Mitra is an irrigation economist at the Gokhale Institute of Politics and Economics which is known for its long tradition of research in economics of irrigation. He apprises the reader with his own thinking on the problems of developing major irrigation in India. Expressing forcefully his reservations about this writer's perceptions, he concludes on a note of (1) an urgent need for an overall long-term water policy and (2) the need for adopting holistic or multi-disciplinary approach to the issues of water resources.

The second comment is from Chandra Shekhar Pant and Shripad Dharmadhikary, both having engineering background. They make some apt observations on the basis of their field experience in the Narmada basin. They draw attention to several shortcomings in the Narmada Valley Project, about which the planners have to do much answering. For instance, they point out that the submergence costs of forests are worked out at ridiculously low prices. They point out that the opposition to big dams is partly because these signify a faulty developmental process adopted in the country, a point made time and again by many critics of big dams. The third comment is by Nirmal Sengupta, a versatile economist known for his valuable contribution to the history of Indian Irrigation development during the colonial era. In his paper, we witness, on the one hand, a 'tearing apart' of the initial paper under discussion. On the other hand, he presents his own loud thinking on Indian irrigation. This writer's brief response to these comments is given at the end of the chapter.

Arun Ghosh, a veteran economist, is the author of the paper 'A Plea for Open Government : Case of Narmada Project' which forms the Chapter 4 in this volume. Utilising the *EPW* column 'From the Ivory Tower', he wrote this paper in July 1989, well before becoming a Member of the newly constituted Planning Commission. He fervently pleads, in the cause of both democracy and right development, for the publication of a white paper on the Narmada Valley Project, so that an open debate on this controversial project could take place. He bemoans the government becoming increasingly insensitive to public criticism.

In a detailed paper Ramaswamy R. Iyer, a distinguished member of the Indian Audit and Accounts Service who retired in 1987 as Secretary to the Ministry of Water Resources, offers a penetrating analysis of the big dam issue. According to him, we cannot say 'no' to large dams on environmental grounds only, nor neglect the importance of capacity utilisation, improved water management, local water harvesting, afforestation and improved groundwater recharge. High priority for minor irrigation works is urged since such works call for less immediate investment, promise quicker results, and create fewer problems. But investments in large dams are imperative. In fact, both large and small dams have their due place in overall scheme of water and land-use planning. But the selection process for large dam projects is what needs to be done with great care and caution. Jasveen Jairath, an upcoming irrigation economist with multi-disciplinary background, offers some major comments on Iyer's paper. Her comments, alongwith his reply, are also incorporated in Chapter 5.

The role of major irrigation works in the Indian economy, as perceived in the upper echelons of water engineers in India, is very comprehensively portrayed in Chapter 6 by R.B. Shah who very recently retired as Chairman of the Central Water Commission. This is perhaps the first significant response by a senior official engineer to the recent criticism of big dams in the country. It first appeared in a little known new engineering magazine (*Civil Engineering and Construction Review*) from Chandigarh in September 1989 issue, but has been republished in the *Journal of Indian Water Resources Society* in its special issue of January 1990 brought out in the memory of late Kanwar Sain, the doyen of irrigation engineers until his death in September 1988. It merits careful perusal, notwithstanding the fact that it is not project specific in responding to the critics of the controversial dams.

The paper by this writer in Chapter 7 was originally written in mid-1988 for the Servants of India Society, Pune, which contemplated bringing out a commemorative volume in honour of Dr. D.R.Gadgil and Vaikunth Mehta. It appeared in *EPW* in September 1989, and some parts of it were carried in my book *Studies in Irrigation and Water Management* which appeared around the same time. Its focus is on the cost side of the controversy between major and minor irrigation works. It points out the questionable nature of the resultant estimates of capital costs per unit area to be benefited by irrigation works when such estimates are derived, plan-wise and irrigation category-wise, from plan documents. In the case of minor irrigation works the derived estimates are grossly understated. In short, the case for according higher priority to minor irrigation in Indian water planning on grounds of cost economy is found untenable. However, Ashok K. Mitra has serious reservations (not yet published in *EPW*).[2] These reservations, alongwith this writer's brief response, are also incorporated in Chapter 7.

Manu Shroff, till recently the editor of a leading economic daily viz. *Economic Times*, is the author of 'Point, Counterpoint' which constitutes the Chapter 8. He had written it as a sort of introduction to several papers on Sardar Sarovar Project, that were carried in a supplement to the *ET* issue of September 28, 1989. He critically looks at some major arguments on both the sides -- Baba Amte, Medha Patkar and B. B. Vohra questioning the big dam idea, and Sanat Mehta, U.K. Srivastava, Anil K. Gupta, Vidyut Joshi, and Christopher Diewald and Thomas A. Blinkhorn stoutly defending it. He thinks the conflict between development and protection of environment is more apparent than real. Perceiving poverty to be the main threat to our environment, he cautions environmental activists not to press for conservation for conservation sake.

2. Dr. A.K. Mitra, on my request, wrote out his reservations which reached me just when the manuscript was being sent to the press. A more detailed response would be prepared later, taking due note of critical observations on the same paper by Satyajit K. Singh in his big paper 'Evaluating Big Dams' published in *EPW* of March 17, 1990.

In Chapter 9, Rusi Engineer, till recently the editor of *Business India,* explores the parameters of the controversy plaguing Sardar Sarovar Project. He finds the authorities that be touting it as Gujarat's life-line. In contrast, the critics condemn it as a paramount case of faulty and perilous development model being pursued by our planners. He finds the critics driven by a powerful new vision of equitable and ecological sustainable development for all, and cautions against the adoption of a dismissive attitude towards them.

In Chapter 10, K.C. Thomas, an ex-Chairman of the Central Water Commission, gives his views about our water resources. The paper was read as the First Kanwar Sain Memorial Lecture in Delhi on 25th January 1990, and later appeared in *Mainstream* of March 17, 1990. He outlines the dynamics of our water resources in quantitative terms, underscoring the imperative of impounding our river flows through massive reservoirs. He strongly questions the alternative of a series of small reservoirs in lieu of a big one from the standpoint of technical feasibility and excessive land submergence. He gives quantitative picture of consequences of reducing dam height of Tehri, Sardar Sarovar and Indira Sagar.[3] He also looks at the hydro power angle of our dams, and concludes on a note of pertinent questions.

The writer's colleague Mrs Kanchan Chopra, a well-established resource economist, offers her thinking on large dams in Chapter 11. She briefly examines the INTACH report on Tehri dam to evaluate its content within the cost-benefit framework. The relevance of cost-benefit analysis as a tool for evaluating large dams is briefly commented upon. Related analytical issues are also discussed, leading to some policy suggestions on water resources development. Finally, in Chapter 12 we present the thinking of a mature, non-official engineer, A. Sundar, under the title 'Evaluation of Public Irrigation System'.[4]

3. Dr. Thomas, at the request of this writer, gathered such data painstakingly from maps and official records about these dams, besides expanding his paper through fresh insertions of three paras.
4. I owe special thanks to Dr. Sundar for complying with my request, amidst a tight schedule, to jot down his views on the raging big dam controversy.

IV

Obviously, the controversy relating to big dams is as unresolved today as it was when it seriously began a few years ago. The discussion during this brief period has only brought out the highly intricate nature of various issues that underlie the problem. In the meantime, significant governmental changes have occurred as a result of the recent general elections both at the central and state levels, owing to which some hope has been generated that authorities would no longer wilfully disregard public criticism. Already, of the two most intensely criticised ongoing projects – Tehri and Narmada Valley – the former has been put under an official review, and the latter might soon follow suit. A climate for discussion and dialogue, and for open exchange of views on wide ranging issues, seems to be forming. While this opportunity lasts, it would be prudent on the part of everyside to this controversy to use it optimally and constructively by paying attention to both positive and negative aspects of every option.

Three general points are of some value in this regard. Firstly, the problem relating to big dam involves many technical issues – seismic, ecological, hydrological, economic, engineering, organizational, etc. – discussion about each of which requires considerable expertise. Therefore, meaningful participation in a discussion of these issues cannot but mainly be limited to experts and specialists. However, there is also a human issue involved, which under no circumstance limits participation in the debate. In particular are those who are most severely affected by these projects. They are a major party to decision making, and ought to be so recognized. No decision can be deemed as acceptable unless it reasonably satisfies their needs and aspirations.

Secondly, it is absolutely naive to entertain the hope that even a most thoroughgoing review by experts can lead to resolution of every technical issue. The truth of the matter is that experts rarely find agreement amongst themselves, and this, alas, is not entirely owing to the imperfect state of the existing knowledge. Therefore, decisions have to be made by the exercise of political will of the society which places reliance on (1) certain major socially shared values and priorities, (2) principle of moderation when interests clash, and (3) preference for safer solutions when technical matters are in dispute, thus not committing the society to perilous or hazardous options howsoever attractive these might seem in the short run.

Finally, there is an aspect which is barely hinted by some critics of big dams, but one which deserves far more discussion and clarity. Their opposition to big dams follows from an overall ideological preference for "small is beautiful". This view may not be shared by all critics of big reservoir-backed ventures either on project specific or other technical grounds. However, it is quite true that no technological options can be made in separation from the vision of the future society that we desire. While the art of planning lies precisely in linking the compulsions of today to a progressive movement towards such a vision, it is indeed doubtful if we have adequately discussed what this vision consists of, let alone evolved some broad agreement about it.

REFERENCES

Dhawan, B.D. (1988). Irrigated versus Dry Farming. *EPW* March 19, 1988

Dhawan, B.D. (1989). Groundwater Development in Low Rainfall Hard Rock Regions, paper for the Workshop on Policies and Management Strategies for Groundwater in Low Rainfall Hard Rock Areas, held at Coimbatore during 4 to 6 October 1989 under the auspices of the Water Technology Centre, Tamil Nadu Agricultural University, Coimbatore.

Dhawan, B.D. (1990a). *Studies in Minor Irrigation*, Commonwealth Publishers, New Delhi (forthcoming).

Dhawan, B.D. (1990b). Groundwater Estimates of India -- A Note of Caution, *EPW* (forthcoming).

2

Mounting Antagonism Towards Big Dams

B D Dhawan

The assumptions underlying irrigation planning are being increasingly questioned. The role of major irrigation works based on big dams in particular is under serious scrutiny by environmentalists, soil conservationists and social and natural scientists. Unfortunately, irrigation planners have baulked at engaging in an open dialogue and discussion with their critics.

Some major assumptions underlying irrigation planning in India are being increasingly questioned in many circles. The idea of big dam in particular has become very controversial, the controversy becoming very serious in respect of the Tehri and Narmada dams. The purpose here is to delineate the main parameters of this rising opposition to the idea of major irrigation works requiring huge reservoirs, and to do some critical stock-taking of the controversy.

While irrigation planners are under attack[1] for approving major irrigation works, they have, unfortunately, tended to ignore this criticism.[2] No wonder their silence and indifference to sharing vital

1. Among the vocal critics who have utilised the popular media indefatigably is B B Vohra, an erstwhile administrator with keen interest in land and water management issues. As an irrigation economist, I have found it an uphill task to bring round otherwise very knowledgeable persons to my point of view whenever my views are at variance with Vohra's standpoint.
2. My friendly advice to some senior planners in both the Planning Commission and the Central Water Commission (CWC) that they should respond to such criticism was not duly heeded. On one occasion, the CWC did respond to Vohra's article in a professional journal of water resources, but the response was ineffectual, as it was a disjointed write-up: separate divisions in the CWC wrote down individual responses which lacked both cohesion and direction.

information with the public has turned many men of public concern receptive to this criticism appearing in the popular media day in and day out. If the planners now feel peeved, they have only to blame themselves (including their erstwhile colleagues) for the spot they now find themselves in. With a formidable line of opponents and public opinion tiltted against big dams, they veritably face an uphill task of defending their position, especially on the recently cleared two big dams on the Narmada river.

Among their persistent critics are both the hard ones and the soft ones. The hard ones comprise the environmentalists, the natural scientists, the social scientists, etc. They raise objections on discipline-specific grounds. But the soft ones reveal a rather woolly understanding of the complex nature of water resources, or of ecological matters, although some of them are persons of eminence in their own right and well imbued with a sense of purpose. In the present situation, they tend to accept uncritically the views of the hard critics. The more the planners have adopted an attitude of dismissing their critics without a reasoned debate with them, the more the latter are pushed to publicising their apprehensions in the popular press. As a result of this one-sided propagation[3] in the media, more and more of the cross-section of our society has taken a credible view of the critics' standpoints.

Just as the planners cannot be right all the time on all matters, so cannot the hard critics. Nor, for that matter, can the soft critics be in the wrong all the time. Who is wrong and who is right on certain issues can emerge through debate and dialogue, both in professional forums and through public media. Here, the planners have double-faulted. Professional forums cannot be meaningfully used if there is disinclination to share information with other professionals. As for the use of the daily press and the weekly magazines, the irrigation planners have mostly shied away from sending to news agencies (like the UNI which has provided ample propagation to communications/memoranda it

3. A recently retired member of the CWC has complained to me that the daily press has not provided them the necessary space for expressing their criticism of the articles/editorials against big dams. He cited the case of *The Times of India* which has carried many write-ups against Tehri dam. Maybe, their rejoinders were too technical, or too near the truth as to make the editorial writers appear foolhardy to discerning readers.

received against big dams) their own version of the controversial points, not to speak of making no effort at all to make specific responses to critical write-ups appearing in the press day after day.

ENVIRONMENTAL ANGLE

Why irrigation planners are disinclined to a public dialogue on issues of public concern we need not pursue here.[4] Let us first have some clearer understanding of the various grounds on which the idea of big dams is being questioned in India. We may begin with the environmentalists who constitute the vanguard of the critics. As I see it, their main worry is on two environmental counts.[5]

First, as the forest cover of the earth is receding at an ominous rate, they naturally worry about any development that causes deforestation. A dam can reduce forest cover in two ways. One, the reservoir behind a dam can submerge forest on the upstream side because the submerged area for a dam can typically vary from 5 to 15 per cent of the irrigation potential created by irrigation project. Two, the dam-building activity can lead to removal of nearby trees for building project roads and housing colonies. As far as the submergence aspect is concerned, the damage on this account is a small proportion of the total damage to forests from all other causes listed below :

(i) illegal felling of forest trees;
(ii) unauthorised ploughing of forest lands;
(iii) authorised allotment of forest lands to dam oustees or the landless people;[6]
(iv) increased pressure from graziers as the village commons and the culturable waste lands are diminishing.

---

4. I can understand that irrigation engineers are temperamentally ill-equipped to write for the benefit of the general public. But this does not apply to Yojana Bhavan whose members do express themselves in public.
5. Many of them are no less worried about two non-environmental concerns: (1) the rehabilitation of the dam oustess, and (2) the excessive siltation of reservoirs that reduces sharply the life of a dam.
6. Allotment of forest land to landless families have been made by state governments, though this is now forbidden under the new Forest Act of 1980; the Centre alone is now empowered to do so. For instance, no less than 2.5 million ha of forest land in Madhya Pradesh state was allotted to the landless for purpose of crop production. *Source* : Varma, p. 101.

According to one knowledgeable dam builder, the total forest area so far lost in India due to submergence by reservoirs created by all big dams amounts to about 5 per cent of the total forest cover we have lost since Independence.[7] Evidently, this underscores an important fact: the problem of our vanishing forest cover is not going to be solved by calling a halt to the construction of big dams. But wherever the forest cover to be lost by a single dam is unduly high, e.g., in the case of Narmada Sagar, it is worthwhile to minimise such loss by reducing the reservoir size via a reduction in the height of the dam. Such 'middle paths', which are well suited to a democratic society, can emerge through wider discussion and sharing of relevant data available with the planners.

The environmentalists' second major apprehension relates to land degradation due to waterlogging and rising soil salinity within canal commands served by big reservoirs. In the light of the historical experience of India, there is no denying that major canals do give rise to problems of waterlogging and salinity. But the statistics quoted in this regard are of questionable magnitude and relevance.[8] Quite often, the cited statistics are for the entire land area suffering from waterlogging and salinity, without any sepearation of the area affected due to canals and due to other causes, natural and man-made. In flat areas receiving high rainfall, lot of land is naturally waterlogged – the problem here is further aggravated when roads, railways and river embankments are built with poor provision for cross drainage. Similarly, in highly arid regions of western and southern India, soil salinity is a natural occurrence which undoubtedly gets aggravated by the introduction of irrigation, particularly canal irrigation.

Indian canal irrigation has significantly contributed to waterlogging and soil salinisation. For this, the blame has been

7. The name of this dam expert is Bharat Singh, a former vice-chancellor of the University of Roorkee which is renowned for its education in water engineering. He delivered a special lecture during the First National Water Convention held in Delhi during November 12 to 14, 1987 under the auspices of the Central Board of Irrigation and Power, New Delhi.

8. Now and then, one comes across the assertion that net irrigation addition from new irrigation works is usually nil because annual increase in waterlogged area happens to match the annual increment in irrigated area through new works. But this is mostly an experience of the arid Middle East countries like Egypt and Iraq, and not of the whole world.

squarely fixed on non-construction of drainage system within canal commands.[9] Thus the problem of waterlogging is not an inevitable consequence of dam building. It could be effectively checked with adequate provisions for drainage, whether through horizontal drains or through vertical drainage via wells and tubewells. In arid regions, the problem of soil salinity is best contained through the use of drip and sprinkler modes[10] of watering crops, because the disposal of drained saline water is an exceedingly tough problem in such areas.[11]

THE SEISMIC ANGLE

Another area of concern relates to damage from earthquake. There is an apprehension that the huge mass of water behind a dam may induce seismic disturbances in the neighbourhood area. Increased seismic activity in the Koyna region has been ascribed by some persons to the Koyna dam. This is an issue which can be best resolved by geologists and seismologists. What is of greater concern is when a dam is proposed on a dam-site situated in an already well-marked seismic belt of the earth, e.g., the proposed Tehri dam in the Garhwal Himalayas. In such cases, a major breach in a dam following an earthquake of severe intensity could be an instrument of utter disaster instead of economic development. Here, too, one has no choice but to go by the judgment of civil engineers as to how much earthquake-proofing can be achieved for dams built in seismic zones. If there is no unanimity among them on the new technology of rock-filled dams, scaling down the dam height would go a long way in lessening the anxiety about damage from dam burst due to an earthquake.

9. One of the main reasons why the drainage component has been missing from our major canal networks is the paucity of investment funds. After all, the cost of drainage network, both in capital and maintenance terms, is no different from that of the canal network itself. It is mainly because of this paucity of funds that even the full complement of the canal system is not completed on many irrigation works even after several years of their being commissioned.
10. Drippers and sprinklers are quite costly to install and operate, not to mention that these cannot be used for all crops. For example, drippers are economical for horticultural farming, while sprinklers cannot be used for irrigating paddy which absorbs almost half the irrigation water in India.
11. Soil scientists, while quoting favourable benefit-cost ratios for reclaiming salinised fields through their research based techniques, never reckon with the problems arising from draining saline waters into rivers that are already being used for irrigation and drinking water purposes.

REHABILITATION OF DAM OUSTEES

Among social scientists, the social anthropologists have never been at ease with one most bothersome aspect of the water related development process--the uprooting of populations from the villages submerged by reservoirs created by damming rivers. Among the affected populations are often sizeable numbers of tribal people[12] whose very survival is already at stake in this rapidly changing world. It is truly a stupendous task to rehabilitate and resettle such dam oustees, and our past record in this vital sphere, thanks partly to our insensibilities to the problems of our tribes, has been very dismal.[13] It is this human question of rehabilitation that is now worrying very many a sensitive soul from various walks of life.

OTHER CRITICS

Finally, one has to reckon with sundry groups and individuals who object to investments in big irrigation schemes for one reason or another. Some are sore as these schemes gobble up, Plan after Plan, the limited cake of investment funds earmarked for the agriculture sector. This has a crowding out effect on other imperative schemes, such as soil conservation and dryland farming technology programmes, thereby incurring hostility from their proponents. While there is need to rectify this imbalance in resource allocation within the agriculture sector, it does not warrant an abandonment of major irrigation

12. The composition of the dam oustees in the four major Narmada Valley Project in Madhva Pradesh alone is as follows:

		Population Affected
1	Narmada Sagar	
	Total	80,500
	Scheduled Tribes	17,653
	Scheduled Castes	11,432
2.	Sardar Sarovar	
	Total	45,000
	Scheduled Tribes	23,240
	Scheduled Castes	3,650
3.	Omkareshwar	12,295
4.	Maheshwar	9,420

13. The government of Gujarat has lately evinced very encouraging response to the demands of the dam oustees and the activist groups fighting on their behalf. According to press reports, over Rs 6,000 million are now earmarked for rehabilitation purpose by it. This would enhance the investment cost of the Sardar Sarovar Project by one-tenth or thereabout.

works programme altogether. If returns from such investments are incommensurate with the promises held out by their promoters, is the ground reality in respect of alternative programmes of soil conservation and dryland farming different?

Then there is a motley group of individuals who are sold out on the philosophy that 'small is beautiful'. They would prefer a river to be tapped by a row-upon-row of small dams instead of one or two big dams. If this turns out to be insufficient for meeting our irrigation needs, they would like to see a well, or a small tubewell, or a farm pond constructed on every farm. Their fervour for the minor irrigation cause is articulated primarily on one or more of the following grounds:

(i) Minor irrigation schemes, especially those based on groundwater, add, acre for acre, much more crop output than do major irrigation works.[14]
(ii) On an area to area basis, the cost of irrigation from minor works is a small fraction[15] of the corresponding cost entailed in establishing major irrigation projects and later developing their command areas.
(iii) Irrigation potential is fully utilised in the case of minor schemes, whereas gross underutilisation of potential arises in the case of major irrigation projects.
(iv) Minor irrigation works are completed in very short time, while the construction of major irrigation projects lingers on for decades. This means quicker returns on investment funds put in minor works.
(v) No environmental threats are posed by minor irrigation works.
(vi) Big contractors, in collusion with irrigation bureaucracy and politicians, have much less scope for siphoning off investment funds in the case of minor than major irrigation schemes.

---

14. Among our minor irrigation works, wells and tubewells are far more important than tanks and ponds -- in the southern peninsula tank irrigation continues to be an important source of minor irrigation. Vohra has been arguing that each hectare under groundwater irrigation is twice as productive as one hectare under major canals. In support, he has repeatedly quoted the researches of this author, but without due regard to all the cautionary notes.
to all the cautionary notes.
15. While J S Kanwar, dryland farming expert, and B. P. C. Sinha, a well known hydrogeologist, think that the magnitude of this fraction is one-fourth, Arun Ghosh, an economist, places it at one-tenth on the basis of the data for the first three years of the Seventh Five-Year Plan.

What is the veracity of each of the above mentioned propositions whose combined thrust is one of highlighting superiority of minor over major irrigation works from every angle? It is a pity that irrigation engineers have never seriously tried to counter these propositions. Being aware of the gaping holes in some of these propositions, they do dismiss them in a lighter vein at their own friendly get-togethers. Since they have shown little inclination to do homework to tell the rest of the world where does major irrigation actually stand *vis-a-vis* minor irrigation, they have failed to stem the rising tide of half-truths and myths about minor irrigation that are being vigorously propagated by their rivals. They may, however, fend off this accusation on the following two pretexts.

TWO PRETEXTS

Their first pretext could be that to carry out comparative analysis in costs or benefits of various types of irrigation works is the economists' job, not theirs. This is valid enough. As the economists in the Yojana Bhavan do have a hand in our irrigation planning, they can, if they so desire, give a helping hand to their engineer planners in the Yojana Bhavan as well as the Central Water Commission to defend themselves from the onslaught of the votaries of minor irrigation cause. For this to happen, the initiative has to come from the engineers, who can also solicit the support of knowledgeable economists outside the portals of the Yojana Bhavan. What has prevented them from doing so? Their cocksureness of the rightness of their own position?

The second plea could be that because of strong linkage between different irrigation sources it is futile to talk of this or that mode of irrigation. To put it differently, minor and major irrigation works are somewhat like two sides of a coin, that is, we need both of them for a balanced development of irrigated farming. Thus it is quite possible that the high benefits from one source of irrigation are partly due to the investment in the other source with apparently lower benefits. For example, in areas poorly endowed with groundwater, good returns on investment in well irrigation are often contingent on replenishment of well water yield by a nearby percolation tank or a canal. Unfortunately, the irrigation planners have not used this line of argument to enlighten their critics or the public.

THE NON-ISSUES

It is imperative to grasp what is the key issue, especially in water short regions of the country. The issue is not of a choice between major and minor irrigation, nor between big canals and groundwater irrigation, nor, for that matter, between wet and dry farming. These are essentially non-issues, partly for reasons of the interlinkages among the different irrigation sources, and partly because of our particular conditions of greater scracity of water than of land resource in large parts of the country. That is to say, we are not well endowed by nature in the matter of our water bounty. So except for water abundant regions like the east Gangetic plains (here groundwater may be a more cost-effective option), we have to tap all sources of water by all technologies if we wish to maximise irrigation cover for our agriculture. Presently, at the national level this cover extends to hardly one-third of our crop area. Even after full exploitation of our water resources through major and minor schemes, this coverage is not likely to go beyond the two-thirds mark. So, it is unwarranted to urge the planners to altogether abandon the Sardar Sarovar project and recommend instead the establishment of one lakh small-scale water holding schemes for the state of Gujarat,[16] which is a water-short state.

The issues of displacement of people by dam construction is certainly not an environmental issue. But it need not be dismissed on the simple plea that environmentalists have no business to take up cudgels on behalf of the dam oustees. Nor should one adopt the questionable stance that the problem is taken care of once rigorous exercises have been carried out in cost-benefit analysis of an investment project. For, one intractable problem of the cost-benefit technique, in spite of so many advances in thinking in this realm, is that of 'whose costs and whose benefits.' If Yojana Bhavan has truly resolved this irresolvable problem, then let us know what methods it has employed to weigh the various costs (psychic, social and financial) of the dam oustees against the benefits to the Gujarat farmers.[17]

It is not simply enough to assert (as a Member of the Planning Commission is reported to have done) that for the Sardar Sarovar

---

16. This is the line adopted by the Narmada Bachao Andolan.
17. In the case of the Narmada Valley Project, the main beneficiary is the state of Gujarat while the non-financial costs would be mainly borne by the people of Madhya Pradesh.

Project extensive cost-benefit analyses have been done and plans to make maximum use of the river water have been evolved.[18] Baba Amte is not making an unreasonable demand when he asks for publication of the benefit-cost analyses pertaining to the Sardar Sarovar clearance by the Yojana Bhavan. Only then we would know how much off the mark are the critics who have tried to demonstrate in their writings that the benefit-cost ratios for the major dams on the Narmada are all well below unity.[19]

Here, one may rightly ask: while maximising the objective of use of Narmada water for irrigation, how has Yojana Bhavan handled other conflicting objectives of river planning? By simply enlarging the storage capacity of a reservoir through an increase in the height of its dam, one very conveniently achieves the objective of maximum harnessing of river flows. But the increase in dam height has disproportionate consequences on the land area submerged under the reservoir. Because of the topography of a river basin, addition to dam height aggravates the land submergence problem in an accelerating fashion. In the particular case of Narmada Sagar, the biggest reservoir in the Narmada Valley Project, the relationship between dam height and area submerged is said to be as shown in Table A.[20]

TABLE A

1.	Dam height (metres)	190	210	230	244	261
2.	Total submerged area (ha)	1000	2000	24000	46000	80340
3.	Incremental area submerged per metre increase in dam height of Narmada Sagar (ha)		50	1100	1571	2020

18. This author had the privilege, thanks to the magnanimity of this Member, to have a look at the BC analysis for the Sardar Sarovar Project as carried out by the Tata Economic Consultancy Services (TECS), Bombay during 1983. If I remember correctly, the nominal BC ratio for this project is 1.34, which shoots up to about 1.85 or so as the TECS appraisers attempt revaluation at social prices, whereby the numerator of the nominal BC ratio is propped up by one-tenth and the denominator lowered by about one-fifth.

19. Claude Alvares and Ramesh Billorey, in their devastating article which damns both the big dams and the Planning Commission, mention for Narmada dams BC ratios which are 0.25 or thereabouts.

20. *Source* : Varma (1985), p. 48.

The corresponding relationship between dam height and land area submerged in respect of other Narmada dams is not known to this author. For Sardar Sarovar, S. C. Varma mentions the fact that in the first stage submergence at the dam height of 106.70 metres (350 feet) the number of affected villages would be 23 (with 84 affected families). This number of affected villages would rise to 182 when the Sardar Sarovar attains its stipulated height of 138.7 metres (455 feet). It thus appears that the problems of displacement of the people, which is now causing great anxiety in many quarters, can be substantially reduced in scale by adjusting the height of the proposed dams on the Narmada.

SUMMARY

Assumptions behind irrigation planning are being increasingly questioned. In particular, the role of major irrigation works based on big dams is under serious scrutiny by environmentalists, soil conservationists, social and natural scientists. Unfortunately, our irrigation planners have baulked at engaging in an open dialogue and discussion with their critics. As a result, more and more cross-sections of the society have become very receptive to the criticism of big dams in the popular media. This is unfortunate. Maybe, a good deal of the criticism is either misplaced or not well founded in facts. But this does not warrant a policy of silence, both to the published critique and to requests from concerned scholars and men of public affairs for relevant information on controversial dams like Narmada Sagar and Sardar Sarovar. In the case of all highly controversial dams, the option of reducing their height needs to be seriously explored by the planners. This way we can accommodate substantially the genuine apprehensions of the critics in the realm of (1) resettlement of dam oustees, (2) loss of forest cover and (3) earthquake-related fears.

REFERENCES

Alvares, Claude and Billorey, Ramesh (1987), 'The Dammed' *Illustrated Weekly of India*, November 1-7

Dhawan, B. D. (1988), 'The Way Out of Narmada Imbroglio', *Economic Times*, December 5.

Ghosh, Arun (1988), 'Irrigation and Water Management', *Economic and Political Weekly*, October 1.

Kanwar, J. S. (1988) (ed), *National Seminar on Water Management: The Key to Developing Agriculture*, Agricole, New Delhi, p 7.

Sinha, B. P. C. (1988), 'Groundwater Development in India: Achievements and Perspectives', in Kanwar (1988).

Varma, S. C. (1985), *Human Resettlement in Lower Narmada Basin,* Narmada Valley Development Authority, Government of Madhya Pradesh, Bhopal.

3

Mounting Antagonism Towards Big Dams

Comments

by Ashok K Mitra

B. D. Dhawan (May 20) has attempted to put the issue relating to large and medium storage route to irrigation development in a proper perspective. While he himself is not convinced with the questions raised by the environmentalists, soil conservationists, social and natural scientists against the role of major irrigation works based on big dams, he has appealed to the irrigation planners to engage in an open dialogue and discussion with their critics. His own defence against the grounds on which the idea of big dams is being questioned, however, does not appear to be too convincing. To begin with, consider first the environmental angle. While he accepts that the forest cover to be lost by a single dam like Narmada Sagar can be unduly high it is no solace to learn from him that damage to forest is much higher on account of illegal felling, unauthorised ploughing of forest lands, authorised allotment of forest land to dam oustees, and increased pressure from the grazier. Besides, for all we know, damage to forest because of all these factors may be the backlash of forest cover lost and people ousted for the construction of big dams. In any case to minimise such losses his suggestion is to decrease the storage height. But, it should not be overlooked that reducing the reservoir size is not merely a technical issue; its economic implications in terms of costs and benefits will have to be worked out in order to examine if the economic feasibility is adversely affected. A more general point, however, is that loss of forest cover cannot be counted only in terms of felled trees, it has to be seen in its entirety in terms of destroying the whole eco-system which in the long run may adversely affect the availability of water in the reservoir through siltation and deficient rainfall in the catchment area. Are all these externalities fully internalised in justifying the projects on economic grounds?

Similarly, the problem of land degradation by waterlogging and soil salinity within the canal commands served by big reservoirs, he suggests, can be effectively checked with adequate provision of drainage or the use of drip or sprinkler modes of water use as the case may be. But, the cost of drainage work is no different from that of the canal network itself and hence because of paucity of investment fund this work is missing from our major canal network. Yet the benefit is estimated on the entire potential created disregarding the fact that owing to lack of drainage gradually in the long run more and more land in the canal command would get degraded. Has this been properly accounted for in *cost benefit analysis* to justify the project on economic grounds?

Secondly, consider the technical issue related to seismic disturbances owing to impounding huge mass of water as in the case of Tehri dam in Garhwal which may cause an earthquake of severe intensity leading to utter disaster. Again, is the safety factor suggested by *Dhawan* considered in full, if so, is the increased cost involved in ensuring such safety economically justify constructing such dams? Tehri dam's case in this respect has been very clearly brought out by Paranjpye which merits very serious consideration.*

Thirdly, the sociological angle of rehabilitation of dam oustees, one of the most sensitive issues, cannot be settled by earmarking a lump sum amount as compensation to be paid as suggested by *Dhawan*. The nature of compensation to be paid, the manner in which it is to be paid, the period over which it is to be paid and whether it covers the annual income earned by the oustees have to be examined in detail. The planners and the official are most secretive about these details. One wonders, if the compensation is arrived at in this manner how far would the project be viable in terms of benefit-cost ratio.

Fourth is the issue directly related to economic aspects of alternatives to large dams. It is very well known that the returns from investment on big irrigation works is abysmally low, project after project,for various reasons, some of which are rooted in the issues discussed above. It would therefore be irrational to be obsessed with

* Vijay Paranjpye, *Evaluating the Tehri Dams*, Studies in Ecology and Sustainable Depelopment-1 An INTACH Series, New Delhi, 1988.

the big dam projects and not to examine threadbare the technical and economic issues related to alternatives, mainly of course, minor irrigation. It would not do to call the assertion of superiority of minor irrigation as 'myths' and 'half truths'; a great deal of homework needs to be done to dispel the myths and half-truths about superiority of minor irrigation, if at all. It is nobody's case that the two are mutually exclusive. To the extent that waterlogging in the canal command could be considerably reduced and that ground water sources could be recharged considerably, under different circumstances, by conjunctive use of surface and canal water there is need for both for a balanced development of irrigated farming.

Lastly, the question of distributive justice is a very relevant consideration in any development project in an underdeveloped country like India. It is because of this consideration that in cost-benefit analysis of irrigation projects one is often confronted with the question of whose benefit at what (whose) cost? Considering the valid accusation that beneficiaries of big dams are mainly the influential and well-to-do farmers and the affected ones are mainly the tribals and other economically weaker sections of the population, a premium ought to be attached to the income distribution effect of a development project which benefits more the weaker sections of the society. Alternatives to big dams like Narmada Sagar and Sardar Sarovar need also to be examined from this point of view.

The controversy regarding big versus small irrigation dams needs to be examined in its entirety in a multidisciplinary approach and not in piecemeal manner as objections raised by environmentalists, ecologists, technicians, economists, sociologists, anthropologists, etc. With the advancement of knowledge and expansion of its frontier it is now possible to analyse and discuss the issues in developmental planning in all its aspects from the long-term perspective so that the social output could be increased by maintaining and preserving the natural re-sources and not at the cost of depleting these. It is against this background that there is an urgent need for an overall long-term national policy for the efficient development and management of land and water resources in the country.

COMMENTS

Chandra Shekhar Pant
Shripad Dharmadhikary

THIS is with reference to the article 'Mounting Antagonism towards Big Dams' by B. D. Dhawan (20th May 1989). In general, the tone adopted by Dhawan is that anybody and everybody is today criticising the large dams, while irrigation engineers and planners are maintaining a total silence on the issue. He also implies that if they really start speaking out, then the arguments of the critics would stand exposed as baseless. Dhawan talks about hard critics and soft ones, saying that the hard critics raise objections on discipline-specific grounds and that the soft ones have only a woolly understanding of the issue. These distinctions of hard and soft are, in fact, an artificial creation of Dhawan's imagination. In fact, not only are the critics raising very important objections on discipline specific grounds, but are going much beyond this. It is these critics who have really understood the complex nature of water resources, ecology, environment and the interrelations of all these to the development process of the country. Through their understanding of the issue in a holistic manner, these critics are able to present an integrated, holistic criticism of the problems of large dams in the socio-political-economic context of our country. It is of course impossible to expect that the irrigation planners, trained as they are to stick to their own disciplines and departments, to even begin to understand this holistic viewpoint. No wonder then, that they, and their spokesman (as Dhawan appears to be), unable to grasp these ideas, are forced to take a defensive stand by calling the critics 'woolly'.

Dhawan (BDD henceforth) says that just as the planners cannot be right all the time, so cannot the hard critics. Now this statement is ridiculous, to say the least. As anyone with a little understanding of logic will point out, what prevents the planners from being right all the time? And how does it follow from this that the hard critics cannot be right all the time? These kinds of statements are just evasive attempts to avoid answering the critics. This same absence of logic can be seen later on in the article which says that 'middle paths' (like reduction in the height of the dam) are well suited to a democracy. One really wonders from where BDD got this strange notion about middle paths being suited to a democracy. How can any path, other than the one that is fully committed in the interests of the masses, be the correct one in a democracy? A democratic society does not permit any compromises on the interests of the masses, and wherever such compromises have been made, it is a defeat for democracy.

BDD says that "why irrigation planners are disinclined to a public dialogue on the issues of public concern we need not pursue here". This statement in fact negates the whole article. We believe that this is the crux of the issue. Unless we explore the reasons behind this silence of the irrigation planners, we shall not be able to understand how these kinds of projects are conceived, planned and executed in our country. And unless we understand this crucial aspect, we shall not be able to see the tremendous vested interests that ensure the continuance of these kinds of projects in spite of the solid body of evidence that has accumulated against them. In fact, the reason why the planners are 'disinclined' to speak out is that in an open, fair debate, not only would these vested interests (like the routine ten per cent, foreign trips, lavish meeting in five star hotels, etc.) be exposed, but they would also be not able to counter the objections of the critics. In fact, they are afraid that the critics would use the planner's data itself to support their own arguments!

BDD calls upon the planners to share their information with other professionals. In fact, we demand this sharing, not as a defence of the planners, but as a part of our fundamental right to know.

In the specific case of the dams on the Narmada, Sardar Sarovar and Narmada Sagar, which engineer or planner would have dared to put the following official informations on public forums?

(1) While arguing their case in front of Narmada Water Disputes Tribunal, each state put exaggerated claims of its ability to make use of the Narmada waters, so that it would be allotted maximum share of these. That is how the Narmada Valley Project now consists of over 3,430 dams, a large number of which do not exist even on paper.

(2) 'Narmada Sagar, as conceived so far...is oversized and has adverse environmental influences disproportionate to benefits derived from it' -- A very very senior planner.

(3) The 75 per cent dependable run off of the river was overestimated by the Tribunal at 27.2 MAF. It is now placed at 23 MAF-- thus the projects were planned with overestimated run off.
(4) The Narmada Sagar dam has been planned without considering the effect (in reducing the water availability) of the dams to be, and already constructed upstream (Tawa, Bargi, Barna and others).
(5) Many critical studies and plans for Sardar Sarovar are still incomplete (or even to be started) -- Command Area Development Plan, Compensatory Afforestation Plan, Rehabilitation Masterplan, Catchment Area Treatment Plan, Problems of Salt Water Ingress and so on..
(6) The silt load, which has a direct effect on the life of the dam, has been underestimated at design at 1.55 ha-m/100 sq km instead of the actual observed 5.62 ha-m/100 sq km.
(7) In the cost benefit calculations, the costs of forests submerged have been considered at Rs. 15 to Rs. 55 per year per hectare! And the life of forests had been assumed at 50 years.
(8) As per the environmental clearance given to the two dams, an important guideline is that the compensatory and penal afforestation should be done in the project impact area. Yet, the compensatory afforestation for Sardar Sarovar is being done in Kutch!
(9) The survival rate for these kind of plantations is not more than 10 per cent. This means that if we really want to compensate the loss of forests, we must plant an area at least 10 times the area submerged -- A senior forest official concerned with the project.

These are only examples of the kind of information that the planners have with them. Is it any wonder then that they are 'disinclined' to reveal it? All the above data has been of course obtained from reliable sources.

We have time and again called for a public debate of the kind that BDD talks about; in fact, at least three times have we sent direct invitations to the Sardar Sarovar authorities, and have called them for public debate. However, there has not been even an acknowledgement from them about this. In one of the letters written to Sanat Mehta, he has answered all the other points of the letter but has conveniently forgotten even to mention anything about the invitation for a public debate. Thus, it is very clear why the planners don't want to speak out, as also what the results would be if they do.

BDD then goes on to some of the issues raised by the critics of these projects. Here also, it is apparent that he starts out with a bias towards the irrigation planners and, in trying to answer the objections of the critics, often contradicts himself. Let us look at the issues raised by him.

Environmental Angle

(i) BDD quotes a knowledgeable source as saying that dams have been responsible for only 5 per cent of the total forest loss since independence. Fair enough. But just because we are unable to control the other 95 per cent of the forest loss does not mean that we should go ahead with this 5 per cent. More importantly, it is necessary to understand that these kinds of dams are an integral part of a development process which is responsible for this 95 per cent of forest loss. And the critics who are objecting these dams are at a more fundamental level also questioning this very process. BDD sadly fails to understand this, and thus can advocate a reduction in the height of the dam as a solution! This of course does not in any way answer the fundamental issues raised by the critics.

(ii) About waterlogging and salinity, BDD first accepts that these are serious problems created by canal irrigation, but then immediately adds that the statistics quoted in this regard are of questionable magnitude and relevance.

One wonders whether BDD would find the following 'of questionable magnitude and relevance' --"out of a total of 42 m ha of area under irrigation, at least 10 m ha is affected by waterlogging and salinity". The source is *Government of India Guidelines for Environmental Impact Assessment of River Valley Projects.*

BDD then talks about proper drainage being able to effectively tackle this problem, but at what cost he does not say. He also tries to say that in arid regions drip and sprinkler irrigation would be the ideal way of containing salinity, but in the footnote says that this cannot be used for all crops, and in any case is too expensive.

Seismic Angle

BDD says that this is an issue that can be best resolved by geologists and seismologists. Nobody denies this. In fact, a lot of geologists themselves are raising this issue. As far as the critics are concerned, they have the full right to quote the arguments of the experts to support their case (as earlier in the article BDD quotes his knowledgeable dam builder). To quote only one such example--from a note prepared by the National Geophysical Research Institute, Hyderabad --"the probability of more frequent earthquakes of smaller magnitudes (compared to 6-6.5) being triggered of by it (Narmada Sagar) would be fairly high".

Rehabilitation Of Dam Oustees

It is very revealing that BDD gives the least amount of space to one of the most vital issues. In fact, though the large scale involuntary displacement of people, especially tribals is by itself a sufficient reason to stop building these kinds of dams, BDD does not make any comment on this aspect except to say that this human question is now worrying many a sensitive soul. That surely is an understatement. In fact, many people, including experts, engineers, social scientists, administrators, anthropologists, concerned citizens, are now coming to the conclusion that proper rehabilitation of such large number of people is just impossible, and these kinds of projects need to be stopped. It is obvious that these kinds of projects are planned and executed by people who have never been displaced in their lives, and who have seen the process of displacement and rehabilitation from behind the safety of their air conditioned offices and their red tape bound files.

Other Critics

It is in this section that the very casual and condescending attitude of BDD towards the critics is fully exposed. His use of words like 'sundry groups', 'some are sore' and 'motley group' betray his lack of grasp of the seriousness of the issues raised by these critics. He talks about the need to 'rectify the imbalance of resource allocation (to large dams)' but 'does not warrant an abandonment of these schemes altogether'.

One wonders whether BDD has the idea of the magnitude of this resource imbalance. "At the beginning of the 7th Plan,there were 181 major and 433 medium ongoing projects with a total spillover of cost of Rs. 26,000 crore. The strategy for the major and medium irrigation sector for the 7th Plan is based on prioritisation of projects with adequate funding for completion of projects which are in advanced stage of implementation..." (Annual Report of the Ministry of Water Resources). This is fair enough, but then why push ahead with the Narmada Valley Project, only two dams of which are expected to cost about Rs. 20,000 crore.

BDD talks about a 'motley group sold on the idea of small is beautiful'. Does he realise that this motley group was led by none other than the Father of the Nation, Mahatma Gandhi, and included individuals like E. F. Schumacher whom the government of India had called in as a consultant to help plan the development of our country? It is unfortunate that the ideas of these men got lost in the glitter of 'modern' technology and model of development that we imported from the west.

BDD talks about the 'gaping holes' in the arguments in favour of minor schemes. We would, of course, welcome any irrigation engineer to come forward and put forth his/her viewpoint on the same, and expose any gaping holes. However, we are clear as to why the irrigation engineers and planners are keeping quite. (We have already talked about these reasons earlier). We challenge the planners to expose the gaping holes in the experiments at Ralegaon Shindi, Shukho Majri, Baliraja Dam on the Yerla and other such schemes.

It should be made clear here that the objections to the large dams are not solely on technological grounds, but also on the more important grounds of social justice, equity and the right of access to resources of the nation to all its citizens. There are two points that BDD makes again and again in his article. One is that the planners are silent as yet, and if they speak out, then many of the critics' arguments would come to naught. We have already talked about this, and to us it is clear that the planners are not speaking out because (a) they simply don't have convincing answers to the critics' argument (b) the vested interests and political pressure do not allow them to speak the truth, for we are convinced that this would only go to support the critics.

Secondly, BDD persistently calls for a reduction in the height of the dams, and implies that this is an effective answer to all the objections raised by the critics. However, it is important to note that those of us who are today opposing the Sardar Sarovar and Narmada Sagar, are doing so on certain very fundamental and basic issues, which simply cannot be satisfied by a lowering of the height of the dam. Can BDD tell us how the lowering of the dam height would fundamentally answer the questions of social injustice in terms of distribution of resources, lack of people's participation and their alienation from the development process of the nation, control and management of the resources by and for the elite few, environmental destruction, involuntary displacement on a large-scale, lack of critical, objective studies by the planning agencies, and a decision-making process in which decisions are based on vested political interests rather than on socio-techno-economic grounds?

Today, the opposition to big dams is aimed at exposing this kind of institutionalised, faulty developmental process which is inherently unjust, and is leading to the destruction of the very resource base on which the survival of mankind is dependent. What we need today is a judicious (water) resource management policy that is technically sound, economically feasible, environmentally sustainable and above all, socially just.

COMMENTS

Nirmal Sengupta

DID B. D. Dhawan ('Mounting Antagonism towards Big Dams', May 20) succeed in putting the issues relating to the controversy around big dams in a proper perspective? No, he did not. It is a very superficial summary of arguments, a very timid appeal to both sides to come to a dialogue. Such appeals may appease friends over tea-table but not work when vested interests are involved. The big dam controversy has attained this height because on the one side are those with tremendous financial and administrative back up, complete lack of accountability, lust for grandiose achievements, unparalleled arrogance and techno-cratic disregard for people. Faced with this strong and unrelenting opposition, the critics had no way but to increase their rank continually with the hope that eventually the rising popular protest would bring down the technocracy. The controversy is very old. It had begun in the fifties. Had the irrigation planners and technocrats paid any attention to these problems at that earlier stage, the critics would not have felt any necessity to appeal to the public. Indeed, they did not, nor did the popular media show much interest then in the criticisms. Dhawan may very well check that the media was far more interested in Nehru's "temples of modern India" than in Kapil Bhattacharya's critical writings, the first appeal against big dams that was made to a wider public. Except for the organ of the Communist Party of India which was leading the movement (yes, those were different times!), no other newspaper had thought it worthwhile to report the popular movements against the construction of the Maithon dam. The senior engineers of Purnea district had quickly perceived the dangerous implications of the World Bank inspired additions to the objectives of the original Kosi project, a brainchild of the Congress Flood Advisory Committee. But these engineers had merely submitted a memorandum to the authorities, for they had no reason to doubt the seriousness of the commitment of patriotic leaders and famous scientists to national development. If that trust is lost today that is because of the misdeeds and misleading pronouncements of those in authority. If the critics have taken to wide scale popular protests it is because memoranda and appeals to good sense have not been attended to. If the popular media are now more eager to propagate the critical views that is not because of their being prejudiced and one-sided. Would not Dhawan consider this as the proper perspective of the debate instead of equating the two sides?

What the planners should have done is now being done by Dhawan. No, I don't mean his rejection of the environmental, the land-degradation and other angles of criticism. If by reading the arguments carefully one reaches a conclusion there is nothing to object to and a set of counter-arguments can be advanced, as Ashok K. Mitra has done (August 5). This healthy debate is welcome. But it would have been proper for Dhawan to limit his review within his expertise instead of freely drawing out engineering alternatives from his hat, like vertical drainage, drip irrigation, scaling down of dam heights, etc., etc. The environmental critics have behaved responsibly by showing, with their respective expertise, the environmental and socio-economic problems accompanying the currently accepted design, while leaving the task of undertaking alternative designs to planners and irrigation engineers who have the right kind of expertise for that. At the most, they have tried to draw the attention of engineers to some possible alternatives. Would not it be more appropriate for Dhawan to appeal to the authority to consider such alternatives seriously and come out with feasibility reports, instead of painting the environmental lobby as being ignorant of simple facts? Just engineering knowledge is not sufficient. For testing the potential of an alternative design a whole alternative project is necessary, complete with topographic and hydrological survey, blueprint, financial support, construction and probably commissioning and evaluation. It is the planners and engineers in authority who alone can command all these resources. But they are adamant not to consider alternatives. Time and money have been spent for grandiose but conventional engineering designs like the Garland Canal only to reach negative feasibility reports. But how many hours and rupees have been spent from any of the state and central government departments to consider alternative designs?

But then social scientists should not think that they can take over the task. It is one thing to suggest confidently an alternative in the pages of *Economic and Political Weekly*, but if Dhawan ever dares to meet any of the officials dealing with the Narmada project they would simply show him that his plan would need as many as 20 medium height dams to substitute for the Narmada Sagar in storage capacity and that there are not many stretches on the Narmada river where one can have a 190 metre or 210 metre fall (i.e.. slope around 1:30). With my scanty knowledge of engineering and hydrology I do understand that the alternatives to the accepted designs are not just as easy to produce as Dhawan feels. If that would have been so, the planners and engineers would have known and accepted them by now; their competence is not in question and they are not necessarily insensitive to or ignorant of social issues. Instead, the apparent indifference and the tightlipped attitude to criticism arise because of the helpless situation they find themselves in while looking for alternatives. There simply is none within the current state of technology. The environmentalists and other critics agree, but they don't suggest quick remedies as Dhawan does. They demand that a moratorium be put on big dams. To the irrigation planners and departmental engineers that is an existential threat, hence unacceptable. May be some day they would realise that there still is a third line of approach. If the current state of irrigation technology would not provide them the necessary help they need look for a radical alternative to that. One just hopes that the authorities arrive at this realisation before the confrontation develops much further.

Dhawan's own line of inquiry would not have been so bad if he would have shown the courage to be 'hard core' and 'persistent'-- the qualities which he despises. New ideas germinate always, but they develop to meaningful ideas only with those who are persistent. I will just concentrate on the single imaginative and fact-based suggestion of Dhawan to show that by pursuing even that line of inquiry one arrives at a radical alternative worth considering. I would guess that Dhawan did not want to side 95 per cent with the anti-dam lobby by permitting only a single medium-sized reservoir with only one-twentieth of the capacity of the Narmada Sagar. If so, his suggested alternative includes 20 medium-sized dams. In other words, he is recommending not 'small is beautiful' but 'medium is beautiful'. The difference between the radicals and moderates lies in the boldness with which they can pursue their thoughts. One can go only up to 'medium', while the other does not hesitate to go beyond. The more radical alternative, pursuing Dhawan's idea, must be to reduce dam heights as much as necessary, even to 2 or 3 metres. There are not many stretches on Narmada allowing a 190 metres fall of water level, but there are thousands of places in that characteristic terrain where these small dams can be built, on the river, on its tributaries and on the diversion canals. Yes, the radicals may succeed where the moderates won't. There is this possibility of having a highly decentralised design or, better, a conjunctive use of centralised and decentralised designs as a potential alternative to the Narmada project.

Unfortunately, there is also a terminological implication of such an alternative and its relative merit is likely to be evaluated on that ground, not on the basis of opportunity cost. In India, a three-metre dam is not a 'dam'; a reservoir with a submerged area of a few hundred hectares is labelled 'minor irrigation'. The engineers, as well as Dhawan, would then reject it summarily saying that they *know* -- presuming it is enough that they alone know -- the 'gaping holes', 'half-truths' and 'myths' about minor irrigation. The moderates are also extremely sensitive to radical presence. If not, why would an esteemed and otherwise polite scholar suddenly begin to throw tantrums the moment he confronts arguments in favour of minor irrigation. Yes, the design suggested here has many open questions and I won't rate it more than an imaginative suggestion worth developing. But the suggestion is not mine. It is Dhawan's except that I did some homework to see if that suggestion of reducing dam heights as a possible ameliorative measure can be made into (at least) a topographically feasible alternative. Probably Dhawan would see more clearly what indeed has prevented the meaningful dialogue that he has now decided to make room for.

Some others may not have such a hangover and may try to explore in details the requirements of that design. If thousands of reservoirs, instead of twenty, are included in the same project but are still spatially well planned in their location, why should not they qualify for being termed 'major irrigation'? At the same time, the thousands of small reservoirs requiring local level management are no doubt 'minor irrigation'. One may feel at a loss. But there is nothing unscientific. The major-minor distinction is a colonial invention. Not everyone follows that unscientific distinction, for every system must be both major and minor, if planned properly. Tamaki argues that ever in canal systems the field levels are the systems, the supply lines are only sub-systems (Akira Tamaki, *The Development Theory of Irrigation Agriculture*, The Institute of Developing Economies, Tokyo, 1977). From the farmers' perspective and for the requirements of agriculture, the minors and field channels are primary, though the administration may find it convenient to consider their area of responsibility, the main system, for themselves. And today, when the need has arisen to plan water allocation for whole watersheds and aquifers, the authority must rise to the occasion to treat whole regions as in a single system and should plan for spatially balanced dispersal of tubewells and minor surface irrigation. If groundwater control regulations are ever proclaimed in India, would not the authorities treat the dispersal over a whole region as a single system? Why not then for minor surface irrigation?

Such are the radical departures necessary if even a rudimentary alternative, as Dhawan suggests, is to be explored with seriousness. The inquiry always leads to such drastic demands even in the scheme of categorisation. Need is felt for a hitherto unknown analytical technique like spatial planning for irrigation sources. Operations and management areas throw up unnerving challenges; viable solutions turn up something like extensive participation by beneficiaries for which minimal achievement is still lacking. And if at all a plan is framed, how to implement it? Who would finance a major risky venture? Where would one get the many personnel who must understand the intricacies of a totally different kind of project? And who knows what are the types of information that must be collected in hydrological and geological surveys for the new kind of projects? And so on. It is a formidable, indeed Himalayan, task. But the fact is that the people are trying. In the course of my travels I have met sometimes such sincere engineers, more often in senior posts, who acknowledge the graveness of the problem and have indeed recommended, or actually tried to implement, one or the other aspects of these multi-pronged tasks associated with alternatives. But the need is so very gigantic that an individual, even if he is a chief engineer, just cannot make any dent. The demand is for changing the whole understanding and the whole of irrigation science.

Unmistakably, a section of irrigation experts have responded to this demand. Irrigation engineering is today at the crossroads and some of the more imaginative engineers are experimenting with alternative designs, improved operations and management, farmers' participation, etc. Their success is still limited, their followers only a handful. To most others they are nothing more than heretics and useless visionaries. But is there any easier way out? If not, why make scientific pretensions by drawing rats out of hats? Let the two serious contenders fight.

Will they succeed? I don't know. I know only that the stakes are far higher than is obvious. Civil engineering opportunities are not unlimited. Once the advantageous locations are exhausted, dam construction is pushed into more and more hostile terrain. With every passing phase it will bring aggravated problems. And then the dams will become old. Officially, they are planned to last about a hundred years. Some may not. Already the older dams are showing signs of the middle-age syndrome, like cracks in embankments and loss of storage capacity due to siltation. From a *cost-benefit analysis,* it will be far more advantageous to dynamite the old ones and construct altogether new dams in new sites. Narmada and Tehri are not culminations; the problem has just begun.

REPLY

B D Dhawan

It is now time to make some brief response to the comments by Ashok K. Mitra (*EPW*, 5th August 1989), Chandrashekhar Pant, Shripad Dharmadhikary (*EPW*, 2-9 September, 1989), and Nirmal Sengupta (*EPW*, 2nd December, 1989) on this author's write-up (*EPW*, 20th May 1989). At the very outset he wishes to record his high appreciation of their attempt to focus attention on an important policy issue by offering their own thinking on the subject, as also their criticism of this author's thinking in this area. It is hoped that the discerning readers of this journal have found the discussion illuminating.

Much of this author's work is mainly in groundwater economics. Still, three considerations impelled him to intervene in the big dam controversy raging in the media columns. First, he was struck by the "radio silence" -- to borrow a phrase from warfare -- being maintained by the senior irrigation planners in responding to their critics in public fora. Despite my nudging at personal plane, they would not bestir themselves - some of them did use the occasion of an inaugural or valedictory address to an irrigation workshop/symposium to make a response of sorts. By writing in the columns of this journal (and in the *Economic Times*), the author had faintly hoped that they would be drawn out of their shells.

Second, he was time and again urged by some of his colleagues and friends, who trusted his irrigation knowledge, to express his thinking on the raging controversy. He soon found that, despite all the planners' claims of sophisticated benefit-cost exercises having been carried out, the height aspect of a dam had not received due attention. As the extent of land submergence by a reservoir is an increasing function of dam height, the problem of dam oustees could be better dealt with by focussing attention on the height aspect of the proposed dam.

Third, he was somewhat appalled by one singular feature of the criticism of big irrigation works, namely, nothing is right with these works while nothing is wrong with minor irrigation works in general and well irrigation in particular. To restate it, the critics were concerned merely with the highlighting of the shortcomings of big irrigation works, and were remarkably quiet in mentioning about the drawbacks of the minor irrigation works. In point of fact, the critical faculty they exhibited in good measure in respect of major irrigation works was conspicuously missing when it came to enumerating the merits of minor irrigation works. In particular, he was struck by their naivete -- whenever some of them chose to speak in favour of minor irrigation (presumably when the compulsions of developing irrigated agriculture in India were grasped by them) -- in asserting things like 'groundwater irrigation is a very low cost option *vis-a-vis* big irrigation works', or 'full utilisation of created irrigation potential is attained in respect of minor irrigation works".

I believe the author's task of a rejoinder has been considerably lightened by the publication of two papers in this journal since May 20, 1989 when the paper under discussion was first published. These are: 'Large Dams : The Right Perspective' by Ramaswamy R. Iyer, and 'Major and Minor Irrigation Works : Cost Aspects of the Controversy' by this author, both published in the Review of Agriculture section of 30th September, 1989 issue of this journal. I commend their careful perusal, both by my critics as well as the reader seriously interested in the big dam issue.

Now, some specific counter comments would be in order. We should be grateful to Sengupta for bringing to our attention that the big dam controversy is not new. He refers to the opposition to the Maithon dam in the fifties, a multipurpose reservoir, the major goal being flood control. But I think the Damodar Valley Project (DVP) has been a success story in the matter of flood protection. About flood protection schemes, we must realise that their very success can enhance the material damage whenever exceptional floods occur. The sense of flood protection afforded by a dam (or an embankment along a river) encourages economic activity in the flood-prone areas. Thus an increase in the level of flood damage before and after the construction of a dam is no pucca evidence of project failure. And one should not forget that (1) the dams in the DVP are deliberately not filled to the full as many habitations in the catchment are not yet moved out for the full submergence level, and (2) the requisite work on the construction of small detention reservoirs did not take place.

Regarding his turning the phrase 'small is beautiful' to 'medium is beautiful', I may clarify that the author was not contemplating replacing one big dam by twenty or so medium dams of equivalent irrigation capability -- that would imply no loss of irrigation capacity. When one scales down the height of a dam for reasons of unbearable costs of land submergence, one is foregoing some irrigation capacity because of the reduced storage. All that the author was envisaging was that such costs outweigh the income gains to be realised from the foregone irrigation potential as a result of the reduced height of the dam. The implicit assumption here is that (a) the weight for 'distributive justice' to be attached to the lost welfare of the dam oustees (for which Mitra makes a poignant plea) on the one hand and (b) the correct valuation of the submerged resources of land and forests (need for which is underlined by Pant and Dharmadhikary) would be high enough to yield a sufficiently sizable number to offset the corresponding number of foregone benefits in the agricultural realm. Can economists help the irrigation planner by providing the scale of weights based on distributive justice? If the planner was fooling us by valuing the loss of forest cover at ridiculously low price (Rs. 15 to 55 per year per ha of forest land), we are not restoring sanity on the matter by asking for a price tag of Rs. 1.25 crore per ha of forest cover -- this number was suggested by the Department of Environment to the project planners, presumably on the basis of some computations carried out at the Forest Research Institute at Dehradun.

Among the innumerable points raised by my critics, many are specific to Narmada and Tehri dams. It is true that I myself had referred to the Narmada dams, but that was primarily to illustrate with concrete numbers how designing a dam of a marginally lower height substantially reduces the consequences of submergence by the reservoir behind a dam. Tehri dam is essentially a hydel project, and not an irrigation project. Mitra commends the perusal of Vijay Pranjpye's evaluation of the same. While I have read this evaluation report (published by INTACH, New Delhi), I have some reservations here.

These reservations stem from the very speculative character of this business of cost-benefit analysis, notwithstanding the scientific trappings it has been accorded over time. (One is struck by the contrast that while serious economists are losing faith in the BC instrument, the non-economists are becoming votaries of the same).

Once a threshold value of BC ratio is declared for project clearance (1.5 in the case of major irrigation projects), project promoters begin fixing numbers and parameters so as to remain above the threshold mark. The speculative nature of the exercise offers ample room for manouvre and manipulation by the ethically-eroded and departmentally-controlled promoters (as well as the opposer). It is time that irrigation projects are evaluated *a la* D R Gadgil's pioneer work of the pre-World War II days. He never offered any BC number in his exposte evaluation, and instead concentrated on sizing up accurately the benefits from the project. Similar attitude towards sizing up the costs would emerge if the insistence on a BC ratio is curbed.

Pant and Dharmadhikary are rightly annoyed that when too many irrigation projects are still incomplete, why start fresh projects like Sardar Sarovar and Narmada Sagar. There is no denying that our track record in completing irrigation works, major as well as medium ones, is dismal -- some of the time and cost overruns appear mind boggling. The problem goes back to the early seventies when too many irrigation schemes were launched, with and without prior clearance with the Centre. That is a blot on Indian water planning, notwithstanding the political compulsions of expanding irrigation quickly in those days (the Nixonian tilt against us in the war with Pakistan on the Bangladesh issue, with dim prospects of a repeat of American aid of generous food shipments of the mid-sixties in the event of domestic crop failure due to droughts). These political compulsions from the international arena got superimposed on the local compulsions of the rural politicians to start irrigation works (on government account) all over a state.

Coming to the alternatives to big dams, we must appreciate the limits in this regard. Irrigation is highly land-augmenting in low and medium rainfall conditions of the tropics. The dryland technology cannot be very land-augmenting here because it is suited to temperate regions where crops have low evapo-transpiration needs. There are limits to groundwater-based irrigation and it is being realised now that this is more of a complementary than an alternative source of irrigation in a "technically sound, economically feasible, environmentally sustainable" water resource management policy that is endorsed by Pant and Dharmadhikary in their concluding observation. Sengupta draws our attention to some tendencies among irrigation engineers, whereby our water resources may be exploited much more imaginatively than has been the case hitherto. "Their success is still limited, their followers only a handful," he observes. May their tribe multiply.

4

A Plea for Open Government

(Case of Narmada Project)

Arun Ghosh

If the present leadership of the country really believes in democracy, in giving power to the people through panchayati raj, if the Planning Commission is sincere about planning for the country's development and not in promoting the sectional interests of a few technicians and contractors, they ought to come out with a white paper on the Narmada Valley Project, give it wide publicity and have it debated openly.

Of late, the science of governance appears to be increasingly oriented to secrecy in regard to the functioning of government, and efficient public relation (or PR for short) seems to have been perfected to an art, to replace open debate and to elicit support for government policies. Thus we find that the government budget today reveals less and less, even as government expenditure gobbles up an increasing share of the national product. The sources of routine, rudimentary data are denied to common citizens, not overtly but through non-availability and inaccessibility. Most official data are "for official use only" ; and at times even elementary facts are held back under the Official Secrets Act.

What is extraordinary about this situation is that the shroud of secrecy extends over even matters concerning vital planning decisions – on projects and programmes that are estimated to cost billions of rupees, and affect the lives and well-being of millions of Indian citizens. The gigantic Narmada Valley Project, encompassing the Sardar Sarovar Project (SSP) and the Narmada Sagar Project (NSP), falls into this category; the entire project has apparently been cleared by the Planning Commission -- though this minor details is never publicised - without the Indian citizenry being informed of either the costs or benefits of this mega project.

There are, of course, many respectable experts who have lent support to this mega project. B D Dhawan, one of the most respected experts on irrigation in India, has extended *conditional support* to the project (vide *Economic Times*, December 5, 1988, and *EPW*, May 20) suggesting that perhaps scaling down the height of the dam (at the SSP) by a mere 35 feet could reduce the area of the submerged land to one-fifth, thereby significantly reducing the number of oustees. He has also suggested that the Planning Commission should make public its cost-benefit calculations.

But then, the government seems to have become increasingly insensitive to public criticism. It is, of course, entirely possible that the project is being pushed through by a group of technologists, bureaucrats and politicians who have a vested interest in getting large projects sanctioned; after all, a large contract always offers the possibility of a sizeable 'cut' from the contractor. But then, *that* would apply to a small coterie, and it is by no means clear why the government as a whole (including the planners and their advisers) have been so tight-lipped on this entire project. Can *anyone* claim with justification that the estimates pertaining to the project should be kept back from the public under the Official Secrets Act ? Or that the doubts of the department of environment -- never officially made known, and yet filtering through somehow--should be suppressed ?

Somehow, the approach and attitude of officialdom on this issue pass comprehension. Indeed, the very credibility of the government becomes open to question, especially when one considers that as far back as September last year, over 500 distinguished citizens from all walks of life had submitted a memorandum to the prime minister, voicing their concern and opposition to this project. If these distinguished and highly respected--each in his or her walk of life-citizens have been wrong, is it not the duty of the government to publicise the correct facts ?

This approach is all the more incomprehensible when one considers all the recent pronouncements of the prime minister, in regard to panchayati raj, in regard to giving power to the people at the grassroots level to determine their own future. How are the approximately 2.5 lakh oustees to be rehabilitated ? Are they being given an earthly chance to determine where they should live, and how? Even assuming that the cost: benefit ratio of the project is highly favourable, can the authorities indicate how precisely the oustees are to be resettled, whether the plans of their resettlement are realistic, whether those already ousted have been rehabilitated ? For, the only information available to the public, in the context of the veil of secrecy surrounding this project, is that the people already displaced from the dam site have so far been totally forgotten; apparently, they being not among those whose lands would be 'submerged' do not even qualify for settlement by the Narmada Valley Project authorities.

As stated earlier, there are no official data to go by ; one has to piece together the costs from diverse sources. Baba Amte has recently cited some figures (vide *Indian Express*, July 9) which appear to have the stamp of authority and accuracy, having been taken from a report prepared by C C Patel and Associates Ltd. (CCPAL). (Incidentally, the name sounds familiar; one C C Patel was once the chairman of the Central Water Commission and later irrigation secretary to the government of India. One does not know whether it is the same C C Patel who has floated CCPAL as a consultancy organisation and, if so, whether there is some nexus between this fact and the speed with which approvals are being accorded by the union government to the project. I would like to add that there is no imputation of any ulterior motive herein; it is just that the orientation, the thinking, the vision of an irrigation engineer is generally in favour of large projects involving the storage of surface water which would otherwise flow down to the sea; such an expert is not able to encompass the *indirect* costs of storage of such water).

But let us examine some of the estimates of CCPAL quoted by Baba Amte. (And I repeat, these are figures as given by Baba Amte in the *Indian Express* cited earlier; they are cited here only because the government does not see the need to release any figures on the project for public consumption). The SSP is likely to cost Rs. 11,154 crore -- one does not know at which reference year prices; the CCPAL report is dated July 1988, so one presumes that the estimates would be at 1987-88 prices. This does not include the cost of the NSP, nor of the many smaller schemes necessary for making full use of the reservoir potential. For financing the SSP, the government of Gujarat would have to provide Rs. 8,500 crore between 1989-90 and 1997-98, i.e., in nine years – apparently, the governments of Madhya Pradesh, Maharashtra and Rajasthan are to fork out the rest – and of this colossal sum, international aid (by the World Bank and Japan) would provide only some Rs. 960 crore.

Baba Amte has given details of the extent of irrigation the project is likely to provide to the drought-stricken areas of Gujarat. Apparently, in Saurashtra, more than 80 per cent of the talukas are not likely to get any water from the SSP; in Kutch 5 out of 9 talukas will not get any water; about two-thirds of the 62 talukas getting water from the SSP are neither drought-prone nor arid; and two-thirds of the 52 drought-prone or arid talukas of Gujarat would get little or no water from the project. The estimate of the union department of environment in regard to the "environmental cost of loss of forests" is placed at Rs. 30,293 crore for the NSP and Rs. 8,190 crore for the SSP.

This piece is *not* intended to be a broadside at the Narmada Valley Project; I do not have with me any facts other than those given out by Baba Amte, and he has argued succinctly against proceeding any further with the project. Even now, the height of the dam can be scaled down-as suggested by Dhawan – so that some 80 per cent of the area proposed to be submerged can be saved from submergence. As a result, the NSP can also be scaled down. That would doubtless affect the economics of the project, but it would save a considerable part of the forest area which would otherwise be submerged. Thus, part of the indirect costs-perhaps not even reckoned in the present cost: benefit analysis – can be avoided. In brief, the entire project needs a fresh look.

But, to reiterate a point made earlier, the present discussion is essentially a plea for more open government. Certainly, there is absolutely no case for such secrecy as exists in regard to major schemes and projects approved by the Planning Commission. That body has doubtless some very competent technologists and advisers; however, these gentlemen do not have a monopoly of wisdom. Being officials, they may be expected to conform; even where they may not agree, they are tongue-tied. But investments of the order involved in the Narmada Valley – of the order of Rs. 20,000 crore in all, if note is taken of all associated schemes – the gestation period involved, the indirect costs in terms of environmental damage and rehabilitation of the displaced persons, all these factors call for a public debate on the *pros* and *cons* of such a project before it is undertaken. It is inconceivable that such an enormous project is taken up without taking the entire nation into confidence, without debating on the complex issues, without comparing the totality of costs with the anticipated benefits. If, the present leadership of the country really believes in democracy, in giving power to the people through panchayati raj, if the Planning Commission members are sincere about planning for the country's development and not in promoting sectional interests of a few technicians and contractors, they ought to come out with a white paper on the Narmada Valley Project, give it wide publicity, and have it debated openly. The members of parliament also need to demand more information on this mega project, for we are committing large resources over a long period of time for this project. Quite apart from anything else, there is obviously little wisdom in making a new start on such a mega project when we have no dearth of important projects which are incomplete and are getting delayed because of the paucity of financial resources with the government. This is a theme which has been emphasised repeatedly in this column earlier; what we need today is to quickly complete projects which are half complete so that their benefits start flowing to the economy quickly. At a time when the cost of such incomplete projects is some Rs. 80,000 crore or more, there is little merit in starting yet another one costing Rs. 20,000 crore, especially when the benefits of such a project can arise only sometime in the next century.

But most importantly, this is, to repeat, a plea for more open government. If the present leadership is sincere about giving power to the people – a shibboleth which is being repeated day in and day out– then the leadership should treat the citizens of this country as responsible adults. And as responsible adults, we have a right to know the full facts on the Narmada Valley Project.

5

Large Dams: The Right Perspective

Ramaswamy R. Iyer

I

INTRODUCTION

The 'big dam' controversy which has been dormant for a while has once again become active. Among both the opponents and the supporters of big dams there is a wide range of persons; the scholarly and thoughtful; the well-meaning but ill-informed; the highly partisan; the populists; and the charlatans. Unfortunately, even at the most distinguished level on either side there seems to be a failure to see the force of the opposite point of view. Most papers and speeches on the subject tend to be partial statements. Anyone who refuses to declare his allegiance to one or the other point of view at the outset may fail to get a proper hearing. Moreover, the general polarisation of attitudes on the environmental aspects prevents an adequate discussion of other important aspects.

Despite that discouraging preamble, the present paper will attempt a discussion of the subject in all its complexity. It will be essentially an examination of issues, both environmental and other, and not a scholarly treatise; and it does not propose to attack or defend specific projects. It is to be hoped that the length of this paper will be found worthwhile, and the occasional statement of the obvious accepted as a necessary part of the effort at comprehensiveness (or an approach to it). Even with this length, the paper will be able to present the numerous issues in no more than a summary fashion; a fuller discussion would have needed a book.

II

'INTERFERENCE WITH NATURE' ARGUMENT

At the very threshold of the discussion, it is necessary to grapple with the guardian at the gates, namely, the 'interference with nature' argument. The argument is that the course of a river represents a natural regime; that the river and its environment including human habitations, tribal settlements, forests, and the entire flora and fauna of the area form a natural ecological system; and that any interference with this is fraught with danger and must be avoided. A moment's reflection will show that this line of argument has much larger implications and ramifications than might appear at first sight, and will take us far beyond an opposition to big dams.

All of human life and development involves an interaction with nature. Humanity, no less than wildlife, is part of nature, and it is the distinguishing characteristic of humankind to modify the environment. The discovery and use of fire, the establishment of human settlements on the banks of rivers, the practice of agriculture : all these constituted 'interferences' with nature. The avoidance of interference with nature is an impossibility.

However, it is obvious that with the advent of industrialisation there was a dramatic change in the kind of impact that humanity could have on the environment; for with industrialisation came an organised assault on natural resources as well as large-scale pollution of the environment. It has been pointed out that the industrial use of timber (in particular for the paper industry) and exports of timber to foreign countries as well as to urban markets within the country led to changes in the pattern of ownership of forest land and to extensive deforestation. It is well known that the forest departments of the state governments have been regarded essentially as 'commercial' or revenue earning departments. Even the famous Forest Research Institute at Dehra Dun was in the past mainly a Timber Research Institute. The protection and conservation of forests is a relatively recent concern. The profligate use of energy (until the 'oil price shocks' changed our attitude to energy consumption) is also a familiar story. By now there is also adequate awareness of the industrial pollution of air and water : the dangers of industrial accidents of which Bhopal and Chernobyl were horrifying examples; the fouling of the atmosphere by automobile exhaust in large cities; the phenomenon of acid rain; the poisoning of the seas and coastal areas because of oilspills from tankers; the depletion of the ozone layer; the problem of disposal of nuclear waste; the warming of the atmosphere and the greenhouse effect; and so on. Even modern agriculture carries with it the dangers arising from fertiliser and pesticide residues in the soil and in agricultural produce; and insecticides and chemicals find their way into dairy and meat products. Not merely forest-cover but the entire environment is threatened, and it is threatened not merely by dams but by the whole course of development that western civilisation (and following it, the rest of the world) has adopted -- which first makes heavy inroads into natural resources of all kinds, and then inflicts grave damage on those very natural resources and the air that we breathe and the water that we drink. However, while there is a groundswell of concern about environmental conservation and protection and a demand for more stringent pollution control standards and the stricter regulation of industry, there is no general call for the banning of industry (except for some opposition to nuclear installations) or of large-scale commercial agriculture. Why then is there a campaign for banning dams and reservoirs ?

Large dams are not a separate phenomenon but a part of the kind of developmental process that has come to prevail. If we attack dams and reservoirs on the ground of interference with nature, we shall have to go further and attack the totality of the kind of demand-based, demand-multiplying, technology-driven, 'growth'-oriented development that is relentlessly marching forward, and turn to modest ways of living which make much less of a demand on natural resources and have a relatively low impact on the environment. There may indeed be a case for such a change, but it is not clear that the opponents of large dams are in fact willing to accept this. This paper does not propose to argue either for 'modern' ideas of development or for the Gandhian alternative; the point that is being made is merely that having chosen certain modes of development it will be inconsistent for us to single out one element or component for condemnation.

Nor is it correct to assume that in the absence of a particular project the environment would have remained undisturbed: much of the denudation of the forest cover takes place for reasons not connected with dams and reservoirs. In fact, in many cases the 'forest lands' which fall within the submergence area of a project have already been so degraded as not to merit the description. Indeed it is possible that such projects, if well conceived and executed, may even have a benign impact on the environment. They facilitate the provision of finance and organisation for essential measures which may eventually lead to a situation better than what would have come to pass in the absence of the projects; and they may raise income levels and generate secondary occupations, and reduce the need to cut trees for fuel or fodder. The current opposition to large dams and reservoirs stems from bitter past experience in relation to many such projects which were marked by an exclusive concern with engineering structures and a neglect of human and environmental aspects; but this need not be an inescapable characteristic of such projects.

A general call for a ban on dams and reservoirs on the ground of interference with nature cannot be sustained. We have to look at the extent of interference with nature in each case and decide whether it is so great as to be totally unacceptable, or whether it is acceptable subject to a number of measures being taken for mitigating if not completely offsetting the impact. It may be argued that the attacks are in fact on specific projects, e.g., Silent Valley, Tehri, Sardar Sarovar, Narmada Sagar; but in many of the articles and speeches on these projects there is an undercurrent of hostility to big dams *per se*; and there are certainly papers, reports and articles which explicitly call for a ban on big dams.

What is needed is that a high-level interdisciplinary group of professionals (including some or all of the following: dam engineers, specialists in earthquake studies, groundwater specialists, agricultural specialists, soil scientists, experts in land use and water use, environmentalists and ecologists, sociologists, economists, and others, as needed) should make fairly detailed preliminary studies which would eliminate all those sites which are likely to be totally unacceptable. (Such an exercise will fail completely if the engineers were to resist stoutly the elimination of any site or if the environmentalists were to insist on ruling out *all* the proposed sites. We have to assume that the exercise will be undertaken with openness of mind on the part of all.) This will ensure that at least in future a good deal of time and effort will not be wasted on detailed project formulation with reference to sites against which there are fundamental objections. This will then leave a certain number of sites for which the preparation of detailed projects can be undertaken. These will of course have to be subjected to a thorough examination from all points of view; such an examination may again result in a negative decision on some projects, but there may be others which can be approved subject to certain conditions.

III

CASE FOR LARGE DAMS

A variant of the environmentalist opposition to large dams is the seemingly thoughtful suggestion that we do not really need them. So far as the present writer is aware, this proposition has not really been argued out in detail. It can mean one of two things. The first is that the same purpose can be served through a large number of small projects. This, however, is not necessarily true. If the same quantum of water as in a large reservoir has to be obtained from a large number of small storages, the total surface area to be submerged may, in fact, be larger as each would be a shallow reservoir; for the same reason, the loss of water by evaporation will be much greater. In any case, that is a hypothetical proposition; not all large projects can be broken down into a number of smaller projects. Water-impoundment is location-specific, and if we reject a favourable location which permits the impoundment of a large body of water, we cannot be sure of finding a large number of locations for smaller storages which together will impound the same quantity of water. Moreover, small dams can be built only on small streams or tributaries. It is the large rivers which carry the bulk of the annual flows and any storages on the main rivers will necessarily be in the nature of large projects. There is no either/or between large and small projects; there is, or ought to be, a complementarity among such projects in an overall plan for basin development.

Further, a large number of small storages with low heads of water may not be able to generate the same quantum of hydroelectric power as one single large storage with a high head of water; and some of the power-generating capacity has to be of the hydroelectric kind. (Alternatives such as thermal generation or nuclear power also create environmental problems of different kinds; and the use of coal, oil or gas involves the depletion of finite resources while hydroelectric generation utilises a renewable resource. In any case, where the possibility of hydroelectric generation exists, it is the best means of providing peaking stations). While there is a place in overall power planning for low-head generation and run-of-the river schemes, they will be subject to the seasonality of the flows and will not have the dependability and resilience which only large storages can provide.

The second, and more plausible, meaning of the statement that we can do without large projects is that we can manage with a combination of other measures such as local water conservation *in situ,* water harvesting, the use of groundwater, greater emphasis on rainfed agriculture, economy in water use, better water management, and so on. While all these things need to be done, it does not follow that they will be adequate for meeting the projected future demand for water.

It has always been an accepted proposition that the vulnerable dependence of Indian agriculture on the vagaries of the monsoons must be reduced through the storage of water for irrigation. The disenchantment with canal irrigation which has set in because of the injudicious use of water and its ill-effects cannot undermine the validity of that basic proposition. Certainly, dryland farming practices need to be greatly improved and the productivity of rainfed agriculture stepped up; but it does not follow that irrigation has ceased to be important. If food production in this country has gone up from 50 million tonnes in the fifties to the present level of 170 million tonnes, irrigation has surely been one of the factors responsible; without irrigation, fertilisers and high-yielding varieties of seeds could not have been used. Irrigation will continue to be needed if the production of foodgrains is to reach a level of 240 million tonnes in the early years of the next century to feed a projected population of 1000 million; and it will have to be a mix of different kinds of irrigation, including canal irrigation based on major projects.

Even the argument that minor irrigation needs greater attention, while it has much force (and we shall revert to this later), does not warrant the conclusion that it is a substitute for major projects. A great deal of what we call minor irrigation is linked to the use of groundwater. In the absence of large reservoirs, minor irrigation will depend even more heavily on groundwater (while at the same time the recharge provided by seepage from canal irrigation will diminish). Apart from the energy requirements for lifting water, this could lead to an over-exploitation of groundwater, which cannot be sustained for any length of time. Recent experience has shown that under prolonged drought conditions surface water tanks and even tubewells fail, and that one has to depend on the large water storages to maintain a modicum of supply. If the production of foodgrains during the drought years did not decline to disastrously low levels, it was only because of irrigation from large projects.

It has for long been urged that there should be a 'conjunctive' use of groundwater and surface water. This does not mean merely the use of groundwater to supplement surface water or vice versa; it means the integrated planning of the development and use of the totality of water resources, including atmospheric water, surface water and groundwater. In this understanding of the concept of 'conjunctive' use, the augmentation of the availability of surface water through large projects is as essential and relevant as the augmentation of groundwater resources through an improved knowledge of such resources.

Industries need water too, for both process and cooling purposes and for townships; so do municipal systems, whether rural or urban, for supplying water for drinking, washing and waste-disposal purposes and for fire-fighting. Projections of total water demand for all uses (agricultural, industrial, municipal, etc.) are somewhat conjectural, but such figures as are available seem to indicate that the future is going to be very difficult.

Various estimates have been made of the total water resources of the country by diverse agencies. The figures in the most recent document, namely, the brochure entitled "Water Resources of India" by the Central Water Commission (April, 1988) are as follows:

1. Nationally available surface water resources	1880 km^3
2. Actually utilisable surface water resources (i.e., the quantum that can be actually utilised having regard to topographic, locational and other constraints)	690.309 km^3

3.	Actually utilisable ground-water resources	418.540 km^3
4.	Total utilisable water resources (2+3)	1108.849 km^3
	Say	1110.000 km^3

As against these figures of availability projections of total annual water demand for all uses are as follows:

2000 A.D.	750 km^3
2025 A.D.	1050 km^3

It will be seen that practically the entire utilisable water resources of the country will be actually needed by 2025 AD. We are, therefore, likely to experience serious water shortages as the years go by. In fact the situation may well be even more serious than what the above figures indicate, as the demand projections seem likely to be underestimates, and the figures of 'utilisable resources' do not refer to resources which are *currently* utilisable, but only what is *potentially* utilisable if all the possible storages and transfers are built. In other words, the figure of utilisable resources *assumes* the undertaking of a number of large projects. (As against the total annual flow of 1850 km^3, storages built or under construction have only a total gross storage capacity of 275.948 km^3 and a live storage capacity of 222.072 km^3).

One may have serious doubts regarding some of these figures, but water scarcity is a reality; a present reality, not merely a future possibility. Already severe shortages are being experienced in most major cities and the hardship is acute in some. Delhi, even though it is close to the Yamuna, has to make a heavy draft on the Ravi-Beas waters through the Bhakra system, and is hoping for an earmarked allocation from the Tehri storage when it is built. Bombay draws water from the Tansa, Vaitarna and Bhatsa storages. Some cities (e.g., Madras, Bangalore) are awaiting the completion of major projects (Telugu Ganga, Kavery Stage-III) to ease their extremely difficult situation; and in one (Madras) the exploitation of groundwater for meeting the city's water-supply needs is reaching dangerous levels. There have been occasions when water has had to be carried to Madras through trains and tankers. The situation in general is likely to get worse rather than better.

That picture of mounting demand and severe scarcity certainly points to a need for demand management. Economy in water use needs to be enforced, wastage penalised, the scarcity value of the resource clearly signalled through proper pricing, technological innovations undertaken for reducing water requirements whether for industry or for agriculture, and the re-use of once-used water promoted. However, these measures by themselves cannot bridge the gap between supply and demand; an augmentation of supplies will also be necessary. As the ultimate source of supply, namely, precipitation, cannot be increased, augmentation means mainly three things: (i) extending our knowledge of the country's groundwater resources and making more of those resources usable through processes such as reverse osmosis; (ii) extensive programmes of re-greening to retard the rate of run-off and improve the retention of water in the ground, as also of water-harvesting and conservation in *situ* ; and (iii) bringing more of the country's notionally available surface water resources into the category of utilisable resources through the building of some large storages (i.e., dams and reservoirs) where these are techno-economically feasible. We have to bear in mind the marked seasonality of the flows (particularly in the peninsular rivers and to a lesser extent in the Himalayan rivers), as also the extreme regional variations. Much of the precipitation occurs within a few months of the year, and even within those months the intensity is concentrated on a limited number of days; and while some areas of the country receive abundant supplies, there are vast areas which are water-short or even arid. It is difficult to see how supplies can be carried over from periods of heavy rainfall to the dry months, or how the water needs of certain areas such as Rajasthan, Rayalaseema, the drought-prone districts of Karnataka, and so on, can be met, without some arrangements for large storages and transfers.

The proposition that we can do without large dams really implies that we can manage with less water. However, the demand for water cannot be substantially reduced without reducing demand of all kinds all round. If we have to do with less water, then we may also have to do with less food, less power, less steel, less coal and less of everything; and settle for ways of living which are less demand-intensive. That point has already been made and need not be laboured again.

However, there are indeed certain possibilities of augmenting our water resources to which adequate attention does not seem to have been paid, namely, drawing water from nature's storages, that is, the seas or deep underground aquifers.

Any inquiry regarding the possibility of use of sea water meets with the response that it would be too costly a proposition. The question of costliness is a relative one of weighing the costs against the benefits, or of costs against alternative costs. It does not seem self-evident that the desalination of sea water and its conveyance over long distances would be necessarily found to be a more uneconomic proposition than alternative means of augmenting our water resources. The costs involved would of course have to be recovered through proper charges and this may in fact necessitate the economic pricing of water, which would be a good thing. Moreover, *prima facie* it would appear that drawing water from the sea may not involve major environmental and socio-economic problems. The possibility certainly seems to warrant more purposeful examination than it has received.

As regards deep underground aquifers, the existence of these (in the Gangetic plains) with vast quantities of water, much of it possibly under artesian pressure, has been postulated by some experts, and the consideration of this possibility has been urged by the World Bank. However, there has been a reluctance on our part (for whatever reasons) to undertake a serious examination of this proposition. Some belated beginnings in this direction seem to have been made in recent years.

However, these are merely possibilities which need to be examined in detail. For the present, the answer to the statement that we do not need large dams is that it is not borne out by facts.

IV

ENVIRONMENTAL AND SOCIO-ECONOMIC IMPACT OF LARGE DAMS

Let us now turn to the specifics of the environmental impact of large dams (with special reference to irrigation projects). The term 'environmental impact' is used here in the widest sense. There is a curious view that the displacement of people is not an 'environmental' aspect. Whatever we may call it, it is certainly a very important aspect, and it is difficult to see how anyone can object to this issue being raised by the environmentalists.

The environmental impact of large projects of this kind would include :

- the loss of agricultural and forest land through submergence under the reservoir which is created;
- the displacement of people and livestock by the project, and the loss of occupations; and in particular the hardship caused to the landless and to tribal communities;
- the dislocation of wild life and the possible disappearance of some rare species of flora and fauna;
- the public health problems arising from the large-scale impounding of water (the spread of disease vectors) and possible climatic changes;
- the inherent dangers of large dams (the possibility of cracks and dam-bursts and the consequent inundation of large areas), particularly in regions prone to seismic activity; and the problem of reservoir-induced seismicity;
- the loss of vegetative cover in the upper catchment resulting in excessive run-off and loss of topsoil, leading in turn to the accelerated siltation of the reservoir and the reduction of its useful life (this could be regarded as a case of the impact of the environment on the project); and
- the emergence of water-logging and salinity in the command area of the project after some years of irrigation, leading to valuable agricultural land going out of use.

We must also take note of certain socio-economic changes which irrigation projects bring about, though these cannot be brought within the ambit of the term 'environmental'. Land values rise with the prospect of irrigation; rich farmers tend to buy up the land; as water becomes available it gets used up by the farmers in the head reaches, and water-intensive crops are grown, so that by the time the canal extends to its full length, the tail-enders get little or no water; all this results in the accentuation of the differences between the rich and the poor farmers. Attempts to settle project-affected people in the command area are also fraught with difficulty. There are also instances in which flow irrigation assures water supply to water-intensive cash crops such as sugarcane, while in nearby areas even drinking water is scarce. The easy availability of canal water and the political difficulties of pricing it economically lead to wastefulness in the use of water, and this in turn aggravates the water-logging problems mentioned above. Lastly, doubts have even been expressed about the introduction of irrigation into arid or desert areas, as it changes established ways of living; for instance, the Rajasthan Canal Project may have the effect of converting an essentially nomadic people practising animal husbandry into agriculturists; or alternatively, it may lead (as in fact it has) to the induction of farmers in large numbers from adjoining states, which could lead to social tensions.

Some of these adverse consequences are readily amenable to remedial measures; some are rather more difficult to deal with; and some are fairly intractable. Among the most difficult problems is the finding of land for settling project-affected people. It is equally difficult to find suitable land for compensatory afforestation to offset lost forests. The re-location and re-training of project-affected people (particularly tribal communities which make a living by hunting and gathering) is a daunting task. The treatment of the upper catchment to prevent the loss of topsoil and the carriage of sediment into the reservoir may present problems, if the project were a large one with the upper catchment far away from the reservoir, and possibly in another state or country. Encouraging economy in water-use, preventing water-logging, and ensuring equity in the distribution of water also present formidable challenges. Such challenges can and must be met; there is no need to take the gloomy view that certain adverse consequences and social evils are inherent in the projects, and draw the conclusion either that they must be endured, or that the projects themselves should not be undertaken at all.

As against these direct and indirect environmental and social costs, the benefits would consist in the provision of water to arid or drought-prone areas; irrigation leading to incremental agricultural production, income-generation and secondary occupations; and in the case of hydro-electric projects, the generation of power and its contribution to industrial activity, income-generation and employment, as also the role of hydro-electric power in balancing the overall power generation system. (No reference has been made to the generation of revenues, as water rates in general are so low that no significant revenues are yielded by irrigation projects).

In the light of the foregoing summary of direct and indirect costs and benefits, it would appear that the matter is merely one of weighing the costs against the benefits and approving those projects which pass the test of a certain minimum excess of benefits over costs.

Unfortunately, the matter is far from simple. There are enormous difficulties in calculating both costs and benefits, and there is a curious reluctance to adopt a proper methodology of evaluation.

V

APPRAISAL AND MONITORING OF PROJECTS: PROBLEMS

Under the procedure now in force, the major and medium* irrigation and multi-purpose projects of the state governments are referred to the Central Water Commission for examination and submission to a high-level committee known as the Advisory Committee on Irrigation and Multi-purpose Projects set up by the Planning Commission (commonly referred to as the Technical Advisory Committee or the TAC). For the purposes of the present paper, it is not necessary to go into the details of this procedure. Nor need we concern ourselves with certain questions which have been raised in this context, such as the need for the central clearance of state projects, the time taken by the processes of scrutiny and clearance, and so on. Considering the massive outlays involved, the need for carefully appraised investment decisions is quite clear, whether this takes place at the central or the state level, and whatever the procedures. What we are concerned with is the quality of the appraisal and the criteria for the approval of projects.

The first point to note is that the engineering point of view is predominant in both the formulation and the examination of large dam projects. A project has its origin in an engineering conception with reference to a location which seems convenient for large-scale water-impoundment, and then the utilisation of the water which the project will make available is considered. The sequence is from project to utilisation, and not from identified needs to a project to meet those needs. To some extent this is understandable, as large-scale water-impoundment must necessarily be planned with reference to a favourable location. However, it is surely necessary to provide some basic justification for the proposed project as the optimal answer to the needs of the area in question and of the basin in which it is situated

---

* The classification of irrigation projects into major, medium and minor in government documents is based on the extent of the culturable command area covered—major: having more than 10000 hectares of CCA; medium: 2000 to 10000 hectares of CCA; minor : below 2000 hectares of CCA.

(unless the project involves an inter-basin transfer, which again will need justification with reference to need and possible solutions). We have to ask whether, from the point of view of the planning of land-use and water-use in the basin as a whole, and having regard to the socio-cultural history of the people, the particular area in question is suitable for agriculture or some other activity; whether it should be rainfed agriculture or irrigated agriculture; what the cropping patterns should be with reference to agro-economic considerations and soil characteristics; what kind of irrigation that cropping pattern will need, i.e., what combination of groundwater and surface water, and whether techniques such as sprinkler or drip irrigation will be appropriate; and in the light of the answers to these questions, what the water needs of the area are and what options are available for meeting those needs. (The consideration of options should include water-harvesting, small tanks, the use of ground water, and so on). If within such a framework we reach the conclusion that a large reservoir and canals are the best answer to the needs in a given situation, there are once again some choices to be considered: the precise location of the dam, the height of the dam, different levels of submergence with corresponding differences in damages and benefits, and so on. Such an examination of basic justification and possible options is not presented to the TAC; what is placed before the committee is only a single engineering proposition (including in some cases some relatively minor engineering choices) for approval. Efforts to introduce an approach of the kind outlined above have not been very successful.

Secondly, the manner in which even the specifically proposed projects are appraised for approval leaves a great deal to be desired. The criterion is conformity to an acceptable benefit-cost-ratio. The basic norm is a ratio of 1.5:1, but there are more liberal standards for drought-prone areas, tribal areas, and so on. Questions can be raised both about the adequacy of the basic norm and about the wisdom of lowering the norm still further for certain areas. However, leaving such issues aside, what is particularly unsatisfactory is not so much the standards adopted as the manner in which the benefit-cost-ratio (BCR) is calculated.

The BCR as now calculated is based essentially on the direct project cost and the value of the incremental agricultural production. (There is a host of difficulties, methodological and other, relating to the very concept of the 'irrigation potential' created by an irrigation project, which is the starting point for the calculation of benefits; however, that is a complicated and perplexing question which cannot be gone into in this paper). The point is that neither the project cost nor the value of incremental agricultural production is a reliable figure.

It ought to be possible to make reasonably firm project cost estimates, given the experience available in this country, but evidently this is not so, judging from the practically open-ended time and cost over-runs to which most major irrigation projects are subject. As regards the benefits by way of incremental agricultural production, these are somewhat conjectural, as no one can say with any confidence that the cropping pattern projected will in fact materialise (not to mention the difficulties of valuation). The possibility that the proposers of the project might have tried to adjust the calculations to yield an acceptable BCR cannot be ruled out. This is a very shaky basis indeed for decision-making.

So much for what the reports present; the omissions are even more egregious. There is very little effort to estimate the socio-economic and environmental costs and benefits. The value of the land needed for the project is taken at the estimated amount of compensation to be paid, and a very rough-and-ready estimate of the costs of rehabilitation of project-affected persons is included. No attempt was being made until recently to value the loss of forest cover or of occupations. Some efforts at such calculations have begun to be made under pressure from the ministry of environment, but they are very unsatisfactory. It should not be unduly difficult to estimate and include measures for countering the public health consequences of water-impoundment, but the provisions made are usually ad hoc and rough-and-ready, and are inadequate. The cost of treatment of the upper catchment is not generally included in project costs, except in a few cases at the instance of the environment ministry or of aid-giving agencies such as the World Bank. (There is a controversy as to whether in fact this is a legitimate charge on the project cost.) Water distribution and the development of the command area will call for further investments before the expected benefits can be realised, but until recently these did not enter the project cost estimates; 'command area development' was thought of as a separate activity. A few years ago, an integrated approach to project formulation covering everything from catchment treatment to command are development was formally adopted, but this has not yet become fully operational. On the benefits side also, there is little attempt to reckon the secondary and indirect benefits.

Industrial, mining and commercial projects in the central public sector are placed before the Public Investment Board after an appraisal by the project appraisal division of the Planning Commission, which, though not a full-scale socio-economic cost-benefit analysis, is a partial approach to it. No such appraisal is attempted in the case of irrigation projects, though these involve massive investments. Some years ago, a committee set up by the Planning Commission brought out a report outlining a methodology for the economic appraisal of such projects. Perhaps that methodology does not go far enough, but it will be a great improvement on the kind of examination that takes places at present. However, it has not been found feasible to adopt that methodology. There has in fact been considerable resistance to such ideas on the part of those involved in the formulation and examination of such projects.

In fairness, it must be admitted that any attempt at a proper socio-economic appraisal of such projects would encounter serious difficulties. The problematic nature of estimating (and valuing) the incremental agricultural production has already been mentioned. The estimation of secondary benefits is still more difficult. It will be relatively less difficult to evaluate the benefits of power generation, if the project were a hydro-electric or a multi-purpose project. (Even in such cases, the quantification of the benefits derived from adding a peaking station to the system is not easy; but this is at least attempted -- separately -- by the central electricity authority.) If the project includes flood-moderation as one of its purposes, the valuation of this benefit also presents problems. On the costs side, it is extraordinarily difficult to quantify the environmental costs. While it may not be too difficult to put a value on the loss of trees or forest produce, how does one quantify the general environmental consequences of the disappearance of forests ? The ministry of environment has laid down a thumb rule for quantifying this, but its soundness is open to serious doubt. In regard to the impact on flora and fauna, the absence of a data base makes it difficult to evaluate this and formulate remedial measures. Expensive and time-consuming surveys have to be undertaken, and this is best done *pari passu* with the implementation of a project, by making it a part of the project and providing organisation and finance for this activity within the project itself.

There is yet another complication. There are costs imposed by the project in the form of loss of agricultural produce, loss of forests and loss of occupations; and there are other costs involved in compensation payments for land acquired or the provision of alternative land, compensatory afforestation programmes, re-training and rehabilitation measures, the provision of financial assistance and infrastructural facilities in the new resettlement areas, and so on. But some of these costs are in fact intended to counter others, and therefore the two work in opposite directions. It will be rather misleading to aggregate all of them; if we do so we must also take into account the resulting benefits, such as the new occupations and the new forests which will arise in due course. In so far as agricultural land is concerned, we can either take the compensation paid as a measure of the loss of agricultural produce and occupations, or we can attempt to estimate the latter in true economic terms; a simple aggregation of the compensation payment and the value of the loss of agricultural produce and occupations would be methodologically unsound. Again, as mentioned earlier, it is quite possible that watershed treatment undertaken as part of a project may eventually result in a better situation than might have prevailed without the project. If so, the costs of such treatment should be offset by the benefits derived from it.

A further difficulty is that costs and benefits do not affect the same set of people. To some extent, this may be true of all long-term planning for development : one generation saves and invests and another reaps the benefits of the investment. However, in the case of large-dam projects there is the further complication that even within one generation one set of people undergoes the trauma of loss of familiar occupations and displacement from areas they have lived in for generations, and others (those who are in the command area of the project or are able to buy land there) derive the benefits of irrigation. In such a situation, how do we weigh the costs against the benefits? Whose costs against whose benefits ? The only answer to this is that the rehabilitation measures should be inspired by the aim of not merely mitigating hardship but also assuring to the affected people a quality of life as good as if not better than what they are losing; and this should be a charge on the project. Attempts have also to be made to ensure an equitable distribution of the benefits arising from the extension of irrigation, if necessary through legislation enabling the allotment of land in the command area to the project-affected persons.

Clearly, then a proper socio-economic cost-benefit evaluation of a major irrigation (or multi-purpose) project is an enormously complex undertaking. It is true that some agencies outside the government have attempted this in the case of certain projects, but their calculations have been questioned by others. However, to recognise the difficulties is not to conclude that the exercise cannot or need not be attempted. The best possible efforts in the given circumstances must be made, and what cannot be quantified should be taken into account qualitatively. Without such an attempt there can be no basis for rational decision-making.

Our difficulties are not over with decision-making. Having made all kinds of calculations and approved a project, we may find that we have no way of knowing how the project will actually proceed. Leaving aside the problem of corruption, there is the persistent phenomenon of slippages, modifications, additions, scope-changes and cost over-runs, so that in the end the project as executed may bear little resemblance to the project as originally envisaged at the time of approval. Provisions may exist in the project for various matters such as proper drainage, environmental measures, the full rehabilitation of project-affected people, and so on, but there is no assurance that these provisions will be properly and fully utilised. In general, the human and environmental aspects tend to receive much less attention once the project has been sanctioned. The funds for compensatory afforestation may indeed get spent but the afforestation may not take place. In theory, water-logging and salinity can be avoided but in practice they may not. The project may envisage the 'conjunctive use' of surface water and groundwater, but there is no operative system to ensure this. Water rates are so low as to guarantee the wasteful use of water, so that the full potential of the project may not be realised. Lastly, the maintenance of the assets created does not receive adequate funding or attention, with the result that the benefits begin to decline rapidly. In other words, the costs both direct and indirect are very real (and irreversible) though they are difficult to estimate, but the benefits may elude us.

It is, therefore, clear that a stringent and comprehensive appraisal is not enough; it is equally necessary to establish a strong, well-equipped machinery which will monitor the progress of all aspects of the project; the engineering and construction aspects as well as the rehabilitation, environmental and water-use aspects. A mere reviewing body will not be of much use; if it is to be effective, it must be armed with adequate powers of enforcement. Unfortunately, the existing appraisal machinery is not really adequate, and there is hardly any monitoring or enforcement machinery.

If the Central Water Commission is to undertake a comprehensive techno-socio-economic appraisal of major irrigation projects, it will need to be transformed into a multi-disciplinary body, not merely by the induction of a few persons from different disciplines at the lower levels, but by providing the agricultural, environmental, economic and sociological expertise at the senior levels (including the top level). A major restructuring will be needed to transform what is in effect a central water engineering commission into a central water commission in the real sense. Alternatively, the project appraisal division (PAD) of the Planning Commission which services the public investment board in regard to industrial projects will have to be expanded to undertake the appraisal of irrigation projects as well; and an appraisal by the PAD should supplement the technical examination by the CWC and these two examinations together should form the basis for a clearance by the advisory committee.

The establishment of a separate body to monitor the implementation of a project once approved, and to ensure not only that project implementation in its construction aspects proceeds apace but that there is also equal attention to the environmental and human aspects, has not really been tried. So far as the present writer knows, such a body has been set up only in one case, and there too it has not yet become very effective.

It must be added that given the political and the (misguided) popular pressure for such projects and the vested interests that get built up, it is most unlikely that any machinery will be established at the state level for a rigorous socio-economic cost-benefit analysis of such projects as the basis for an investment decision, or for the effective enforcement of the remedial measures on the human and environmental fronts in the project-implementation stage. Needless to say, under our federal structure, any attempt at the central level to make the appraisal methodology more rigorous or to establish a post-approval monitoring and enforcement machinery will meet with strong resistance from the state governments. It may, therefore, seem to some that recommendations regarding a stringent examination of projects and the strict enforcement of conditions are unrealistic. From this, the drastic conclusion may be drawn that the only solution to the problem is not to go in for such projects at all. However, this too is unrealistic. Whether we like it or not, such projects will continue to be undertaken. We must, therefore, see how best a measure of discipline can be introduced.

Undoubtedly this will be difficult. We have the familiar phenomenon of a large number of 'unapproved' projects, that is, projects which have been taken up by the state governments without central clearance; and even the Planning Commission finds it impossible to refuse to make provisions for such projects. There is also the other phenomenon of 'non-plan projects'. that is, major irrigation projects which cannot be included in the plan and made eligible for central assistance because of non-clearance by the centre, and which are, therefore, being financed by the state governments out of non-plan resources. Nevertheless, in spite of these difficulties, we have no alternative but to keep trying to bring about better discipline in this regard through such means as are available to the centre, namely, the requirement (grudgingly accepted by the state governments) of a scrutiny and clearance for inclusion in the plan, and a legal requirement of central clearance under the Forest Conservation Act, 1980. We must use these means to the fullest extent possible and simultaneously build up the right attitudes on the subject, through an extensive dissemination of information and analyses.

In this context the point that has been made by several writers regarding the need for openness on the part of the project authorities and the governments concerned is very important. Information regarding all aspects of a project should be freely available; studies by academic institutions, consultancy agencies, voluntary bodies and others should be facilitated, and there should be receptivity and not resistance on the part of the project authorities and the governments concerned to the suggestions and ideas emanating from such outside sources.

It will be very useful to undertake an evaluation of a number of completed major projects with a view to ascertaining the nature of the socio-economic and environmental impact that they have actually had, the benefits that they have contributed, and the costs that they have imposed. Such studies, of which there are regrettably very few, would provide invaluable guidance for the consideration of new projects.

VI

INVESTMENT COSTS AND THE FUNDING PROBLEM

We have not so far considered the funding problem. The capital costs of major irrigation projects have been going up sharply over the years, and at the same time the projects take longer and longer to complete, and there is always a heavy backlog of uncompleted projects for which funds have to be provided in each plan. According to the Annual Report of the Ministry of Water Resources for 1988-89, as many as 181 major projects were spilling over from various plan periods into the Seventh Plan, needing a provision of Rs. 26,270 crores. However, the actual provision in the Seventh Plan for major and medium irrigation projects was only Rs. 11,555.56 crores. Clearly, this was inadequate even for the needs of 'ongoing' projects and left hardly any scope for new projects. In fact, the Seventh Plan gave priority to the completion of unfinished irrigation projects which were in an advanced stage and were capable of yielding full or partial benefits in the plan period, particularly those benefiting tribal areas, drought-prone areas and areas with a substantial scheduled caste population; and restricted new starts to medium projects in drought-prone, tribal and backward areas. It also envisaged greater emphasis being laid on minor irrigation programmes which could be completed quickly and start yielding benefits, and the highest priority was to be accorded to measures for the utilisation of the existing irrigation potential. The position regarding provisions to be made in the Eighth Plan is likely to be even more difficult. Priorities will probably have to be still more stringent in the Eighth Plan, and it is difficult to envisage a large provision for new projects. In fact, there will have to be a rigorous review even of the so-called 'ongoing' projects with a view to concentrating on some and completing them; it may even be sensible (though probably politically difficult) to stop work on those projects on which not much progress has been made so far. The position which emerges is that a general debate regarding the desirability of large dams is perhaps somewhat academic, considering that there are hardly any resources for undertaking such projects.

However, this is a question of relative priorities. To rule out major irrigation projects entirely on the grounds of the heavy investment they need and the constraint on resources is to make an implicit priority judgment against these projects and in favour of others in the same sector or other sectors. While giving greater priority to the utilisation of the irrigation potential already created, the reclamation of the potential which has been lost through water-logging and salinity, local water conservation and a vast improvement in water management, we cannot rule out new projects completely. We cannot say that there will be no new major irrigation projects in the Eighth Plan. That would be like the old argument of the sixties in favour of a plan holiday: the answer to that argument is simple, namely, that it cannot be done. The very fact that these are long-gestation projects means that early starts are needed; the uncertainties in regard to costs and benefits and the difficulties of control over the projects are not arguments in favour of abandoning planning, but merely call for greater and more concerted efforts.

Another argument relates to the rate of investment per hectare of irrigation potential created, which has been going up. It was Rs, 1,530 in the First Plan period and has been reckoned at Rs. 39,330 for the Seventh Plan period. (The latter figure has been arrived at with reference to the likely actual outlay of around Rs. 11,800 crore on major and medium irrigation projects and the irrigation potential of 3 million hectares actually expected to be added during the Seventh Plan period). The figures are somewhat misleading because we cannot really compare First Plan figures and Seventh Plan figures without an adjustment for price-level changes. Moreover, there is also a time-lag problem : some of the expenditure incurred during a plan period may fructify into irrigation potential in the next plan period. (But of course, the time-lag factor applies at both the commencement and the conclusion of the plan period, and so perhaps it may not significantly affect the calculations.) To get out of these difficulties, it has been suggested that the investment per hectare of irrigation potential should be determined with reference to individual projects recently cleared by the TAC. One calculations is that this would be Rs. 22,600.

However, this too may not be dependable because the completion costs of the projects might turn out very much higher than the costs approved by the TAC, and the irrigation potential created might turn out to be less. On the other hand, if we try to calculate the cost with reference to a few completed projects, we will run into the difficulty that these projects must have commenced several years ago, and that a substantial part of the investment in these projects relates to past years and will not reflect current costs.

If we ignore all these complications, refrain from making comparisons between different plan periods, and go merely by the Seventh Plan provision of Rs. 11,555.56 crores and the projected creation of additional irrigation potential or 4.3 million hectares, we get a figure of Rs. 26,874 as the investment per hectare of irrigation potential. To this we have to add a figure, say Rs. 2,000 or so (this is purely guesswork), to make up for the likely under-provisioning of environmental and rehabilitation costs; and a further figure of Rs. 2,500 to Rs. 3,000 to cover investments on CAD works, and so on, to ensure the utilisation of the water made available. We then get a rough-and-ready figure of around Rs. 32,000 per hectare. As against this, the Seventh Plan provides an outlay of Rs. 2,804.99 crores on minor irrigation; if we add the projected institutional investment of Rs. 3,500 crores, we get a total of Rs. 6,304.99 crores. This was expected to create an additional potential of 8.6 million hectares. This gives an investment of Rs. 7,331.38 per hectare. Whatever the methodological questions and the adjustments needed, it seems abundantly clear that the investment per hectare of irrigation potential created by a major irrigation project is a multiple (four-fold or more) of the figure under minor irrigation projects.

VII

MAJOR VERSUS MINOR IRRIGATION

It is in this context that the advocacy of a shift of emphasis to minor irrigation acquires plausibility. An economic comparison of the two is of course not very simple. In the first place we may have to make some addition to the investment cost under minor irrigation to cover elements which it does not take into account. Secondly, the cost of operation (energy for pumping, etc.) may be much higher in the case of minor irrigation, and there may be need for capital repairs or renewals and replacements every few years. (Assets created under major irrigation projects have a much longer life.) These are matters for detailed calculations. (The Central Water Commission believes that in an economic comparison the advantage is with the large projects.) The more important point is that, as we have seen earlier, the use of surface water and that of groundwater cannot be considered in isolation from each other.

However, without attempting an economic comparison between major and minor irrigation, it has to be conceded that there are some practical advantages in the latter. In a capital-scarce situation, where the resources to undertake and complete large projects are not easy to find, we may necessarily have to give some preference to options which call for limited investments. Further, we have a much greater control over small projects. These are not plagued by time and cost over-runs. The area covered is limited and so there should be no time-lag in utilisation; the gap between potential created and potential utilised is essentially a problem relating to major irrigation projects (though it has begun to make its appearance in the case of minor irrigation also). The question of a conflict between head-reach farmers and tail-enders is also one which does not arise in the case of minor irrigation. Minor irrigation is also unlikely to create problems of water-logging and salinity. As the farmer ultimately bears both the capital and the operating costs under minor irrigation (though with some subsidisation), water-use is likely to be much more careful and economical than under major irrigation. Lastly, major irrigation projects have a long gestation period whereas minor irrigation yields immediate returns. In view of these considerations there is certainly a case for a much greater emphasis being placed on minor irrigation, and for being more selective about major irrigation, though this does not mean a ruling out of major projects. In fact we need a change of names: the terminology of 'major' and 'minor' carries an unintended implication of a ranking of the two kinds of irrigation. It is necessary to remember that what we call 'minor irrigation' is minor only in respect of area covered and not minor in importance.

VIII

PROJECTS TO BE MULTI-PURPOSE

Finally, considering the heavy investments, long-gestation periods and the human, social and environmental problems involved in large-dam projects, it stands to reason that every possible benefit should be extracted from such projects; in other words, they should be multi-purpose projects. Each such project should provide for drinking water, irrigation, industrial use, the generation of hydro-electric power and flood control (if the river is prone to floods), and the reservoir should also be used for fisheries development. Additional investments (power station and distribution network, water supply systems, cold storage and transportation facilities, etc.) would of course be needed for the actual realisation of some of these benefits. Some of these investments may be a part of the project from the beginning, and others may be made later and may be separate projects, but the water-resource project should be conceived of as a composite one from the start. The economics of that project would undoubtedly be greatly enhanced by such an approach. (Needless to say, each such project should be conceived of, not as an isolated project, but as one component in an overall plan for the development of the basin as a whole.)

IX

CONCLUSION

The purpose of undertaking this rather protracted examination was not merely an academic clarification of issues but also a practical one of deriving guidance for policy and action. The results of this analysis can be set forth as follows.

We cannot, on environmental grounds, say 'No' to large dams and reservoirs; nor can we, having regard to projections of demand and availability, accept the view that there is no need for such projects. We should certainly accord priority to the utilisation of the potential already created, the reclamation of the potential which has been lost through misuse, and a vast improvement in water management (including both economy-in use and recycling). We must encourage extensive local water-harvesting, and undertake re-greening and other measures to retard the rate of run-off and improve the retention of water in the ground. We should also also place a much greater emphasis than in the past on minor irrigation, which calls for less immediate investment, promises quicker results, and presents fewer problems. Possibilities such as the use of sea water and the tapping of deep underground aquifers also need to be examined. However, we cannot rule out investment in at least some large-dam projects. Large and small projects, and the use of surface water and groundwater, have to be integral parts of an overall plan of land-use and water-use for a drainage basin or watercourse system as a whole. At the same time, considering the heavy costs (financial, human, social and environmental) involved in large-dam projects, we have to be highly selective and extremely cautious regarding approvals to such projects. First, there should be an elimination of sites which are totally unacceptable from the environmental or socio-economic points of view. In respect of other locations, strenuous efforts need to be made for a comprehensive and rigorous techno-socio-economic appraisal of large-dam projects, despite the methodological, institutional and political difficulties involved. All available options for a given situation should be examined. Major water-resource projects involving large dams should be considered acceptable only if they represent the optimal answers to the needs of the area. Each such project should be planned from the beginning as a composite multi-purpose project (covering irrigation, hydro-power, drinking water, industrial use, fisheries development, etc.) with an overall development plan for the basin. The clearance of such projects should be conditional not merely on their passing a thoroughgoing scrutiny, but also on the benefits significantly exceeding the costs. Stringent conditions, particularly in regard to the human and environmental aspects, will have to be attached to such clearance, and there should be a machinery armed with adequate powers to enforce the conditions. This will be difficult but it is the only way in which the dangers inherent in such investment decisions can be minimised and the projected benefits realised.

Some of the elements in that statement may seem familiar; they may be found in writings and speeches which advocate different (and partial) points of view. The merit of the present contribution, if it has any, must lie in the openness of mind it strives to bring to bear on the subject; the route of patient examination through which it seeks to reach its results; and the effort it makes (with whatever measure of success) at completeness of statement. The result may please neither the environmentalists (or some among them) who would strongly argue for a simple 'No' to such projects, nor the engineers who may fear that the conditionalities proposed would be so stringent as to be virtually negative; but alas, in this as in many other matters the truth is difficult and complex and does not lend itself to simplification.

[This paper has been written in the author's personal capacity and the views expressed are his own. In the course of the writing of the paper the author received useful information and documents from, or had valuable discussions with, a number of officials, present and former and experts, including G. S. Jakhade, M. S. Rao, Z. Hasan, D. K. Dutt, V. V. R. K. Rao, A. Sekhar, C. G. Desai, B. B. Vohra and Lovraj Kumar. B. G. Verghese, V. V. R. K. Rao, K. S. Sharma, A. Sekhar, G. S. Jakhade and C. G. Desai were also kind enough to read the paper in draft form and offer useful comments. The writer's debt to all of them is gratefully acknowledged; errors of perception, fact or analysis that remain are his own.]

REFERENCES

Government of India Documents

Report of the Irrigation Commission, June 1972.

Report of the National Commission on Agriculture, 1976.

Rashtriya Barh Ayog (National Food Commission) Report, March 1980.

Planning Commission, *Report of the Committee to Review the Existing Criteria for Working Out the Benefit Cost Ratio for Irrigation Projects* (Central Water Commission), February 1983.

Ministry of Irrigation, *Guidelines for the Preparation of Detailed Project Reports of Irrigation and Multipurpose Projects*, 1980.

Ministry of Irrigation, *National Perspectives for Water Resources Development,* August 1980.

Ministry of Environment and Forests, Circular, Submission of Proposals for Dereservation of Reserved Forest or Diversion of Forest Land to Any Non-forest Use under the Forest (Conservation) Act 1980 - Cost Benefit Analysis (No 11-61/85-Fry Cons) dated 13.1.1985.

Ministry of Environment and Forests, *Guidelines for Environment Impact Assessment of River Valley Projects*, January 1985.

Planning Commission, Circular, 'Irrigation and Multipurpose, Flood-Control, Drainage, CAD, Anti-Waterlogging and Anti-Sea Erosion Projects' (16(12)/85-ICAD dated 28.10.1985.)

Ministry of Water Resources, *Towards a National Water Policy: Issues for Consideration*, October 1985.

Planning Commission, *Seventh Five Year Plan*, November 1985.
Ministry of Water Resources, *National Water Policy*, September 9, 1987.
Central Water Commission, *Water Resources of India*, April 1988.
Ministry of Water Resources, *Annual Report 1988-89.*

Proceedings of Seminars, Workshops, etc.

"Indo-Soviet Workshop on Evaluation and Modelling of the Impacts on the Environment of Water Resources Projects", New Delhi 1985.
First National Water Convention, New Delhi, November 12-14, 1987.

Books, Papers, Articles, etc.

Tata Economic Consultancy Services: *Economic Appraisal of Sardar Sarovar Project.* Report prepared for the Narmada Planning Group, May 1983.
Centre for Science and Environment: *State of India's Environment 84-86, Second Citizens' Report* 1985.
B. B. Vohra: *The Greening of India,* INTACH Development Series, No 1, 1985.
N. D. Jayal: 'Eliminating Poverty – Ecological Response', INTACH Development Series 3, 1986.
N. K. Sharma: 'Large Dams: A Necessary Development Choice', *Bhagirath,* Vol 33, April 1986.
B. B. Vohra: 'Some Thoughts on the Conservation of Water', *Aquaworld*, July 1986.
B. S. Bhavani Shanker: 'Comments on B. B. Vohra's Article', August 1986.
R. Goodland: 'Hydro and the Environment: Evaluating the Trade-offs', *Water Power and Dam Construction*, November 1986.
M. G. Padhye: 'Environment vs Water Resources Development', *Times of India*, New Delhi, January 28, 1987.
Sunil K. Roy: 'How Dangerous Projects are Pushed', in *Tehri Dam- A Prescription for Disaster,* INTACH Development Series 6, 1987.
D. K. Dutt, Techno-Economic Consideration for Ground Water Development in India, *Natural Resources Forum,* United Nations, New York, 1988.
Claude Alvares and Ramesh Billorey, *Damming the Narmada- India's Greatest Planned Environmental Disaster*, Natraj Publications, Dehra Dun, 1988.
Vijay Paranjpey. 'Sardar Sarovar Project on the Narmada -- An Economic Folly', INTACH, "Evaluating the Narmada Projects - An Interim Synopsis", India International Centre, October 15, 1988.
G. S. Jakhade: 'Need for Development of Water Resources and Water Conservation', *Journal of Indian Water Resources Society*, Vol IX, No 2, April 1989.
Madhav Gadgil, "Deforestation: Problems and Prospects", Foundation Day Lecture, Society for the Promotion of Wastelands Development, May 12, 1989, New Delhi.
B. D. Dhawan: 'Mounting An tagonism towards Big Dams', *Economic and Political Weekly*, May 20, 1989.
Baba Amte: 'Unquiet Flows the Narmada', *Indian Express*, July 9, 1989.
C. G. Desai: 'Planning Targets for Irrigation Development.'
-- 'Substitutional Irrigation: A Reoriented Planning Concept.'
Baba Amte, 'River Valley Projects and Environment -- The Farce of Conditional Clearance', *Indian Express*, July 13, 1989.
B. B. Vohra, 'For a Bridge over Troubled Waters', *Financial Express*, New Delhi, July 17, 1989.
Arun Ghosh, 'A Plea for Open Government – Case of Narmada Project', *Economic and Political Weekly*, July 29, 1989.
Ashok K. Mitra, 'Mounting Antagonism towards Big Dams', *Economic and Political Weekly*, August 5, 1989.

THE IRRIGATION DEBATE

An Intervention

Jasveen Jairath

Of late a fierce debate on the question of irrigation has been going on in the country ranging from activist forums to popular media right through to academic circles. Stimulated by the popular opposition to the Narmada Valley Project the debate has typically assumed the form of taking positions on the question of large surface irrigation projects in relation to small/medium/underground irrigation schemes. Apparently dealing with technical and specialist matters, the debate has forced issues of a much broader and fundamental nature to the fore, albeit quite unconsciously. The pertinent point about the debate is not so much the particular positions taken as the rationale underlying the respective positions. Essentially it reflects a clash of two distinct viewpoints on the question of development as such. This is explicity acknowledged by those critical of the installation of large dams, who argue that an irrigation policy as above, has to be viewed in the context of an overall strategy of development which is alienated from and superimposed on the people whom it seeks to develop. It is not the material or ecological consequences associated with large dams that they attack *per se*. Their critique rather derives primarily from the social context of the creation and use of large dams which they argue results in peculiar techno-economic consequences. Perforce the tendency of this camp is to widen the terms of debate and discussion. In contrast the focus of the proponents of large dams policy typically remains rather narrow and technocratic.

A welcome exception within this set has been Ramaswamy Iyer's (1989) recent contribution in this journal. He argues the case for qualified support for large dams in as favourable, balanced and rational a manner as possible, unlike some others like Dhawan whose conceptualisation of the issue tends to be more economistic and technologistic. Notwithstanding his support for large dams, Iyer's presentation is quite sensitive to the complexity of the issue at stake.

He is by far most generous in recognising the various externalities associated with large dams and makes an honest attempt to squarely reckon with them. This is in contrast to some of his fellow proponents, who simply evade the issue and indulge in shallow rationalisations of their position that discourage any meaningful dialogue.

Considering his contribution as a representative example of the most favourable presentation of the case for large dams we will critically review it to point out the basic fallacy in the general approach of those whose arguments on the irrigation question are confined to taking positions on small/medium/minor projects or surface/ground water development or even those who propose an integrated water resource development (which may well be a rational course of action). It is the positing of the issues as such mutually exclusive choices between various positions that appears problematic. As a starting point of departure for the ensuing discourse, it lends an entirely different complexion to the discussion the dominant concerns of which are the relative merits of various propositions primarily from a technological-managerial perspective. What follows is a loss of communication between this group and those who have vehemently opposed the construction of large dams, as if, it appears, on grounds of principle. The basic weakness of the above approach lies in its failure to grasp the critique of large dams as fundamentally a critique of an anti-people development trajectory as such. Although technological adversities and management problems with large dams are pointed out as problematic, the attempt there is to focus on the institutional modes of their emergence. These modes derive from the particular (social) conditions of creation and utilisation of such irrigation sources. The attack therefore is essentially on the totality of these social conditions which bears the stamp of an anti-people development perspective. Hence the arguments of people like Iyer who, while recognising the problems with large dams, argue for greater efforts to mitigate them, fail to meet the basic point raised by the opponents of large dams.

To begin with Iyer counters the argument of the critics wherein they oppose large dams on grounds of the ecological disturbances entailed in their construction. Such a mode of reasoning he argues (and correctly so) could well challenge,

> not merely dams but the whole course of development that western civilisation (and following it, the rest of the world) has adopted–which first makes heavy inroads into natural resources of all kinds, and then inflicts grave damage on those very natural resources and the air that we breathe and the water that we drink (Iyer, 1989, p A-107).

In a similar vein further questioning the dominant development paradigm he goes on to argue that, "....if we attack dams and reservoirs we shall have to go further and attack the totality of the kind of demand-based, demand-multiplying technology-driven, growth-oriented development that is relentlessly marching forward...." (Iyer 1989, p A-107). But that is precisely the point – why not go forward ? Why withdraw and merely settle for improving the viability of a strategy of imposed development often at heavy cost to those it seeks to develop. Large dams, as he rightly argues, are an integral component of such a strategy. However, Iyer withdraws in the face of these insights forced upon him during his appraisal of the issue and concentrates more on questions of improving the efficiency of the entire process of decision-making, initiation, execution and implementation of large-scale surface irrigation works. This has correspondingly tame implications for the subsequent discussion on the need for these works and the remedial prescriptions suggested for mitigating the associated adversities.

For instance his arguments on the need-based justification of large dams follows on the pattern of water-needs thrown up by growth-oriented development strategy being pursued. However, juxtaposing absolute quantum of resources against absolute 'needs', following a strongly Malthusian mode of reasoning, abstracts from questions of alternative pattern of utilisation, of needs – quite apart from the question of streamlining wastage, loss, misuse, etc. These alternative 'needs' derive from correspondingly alternative rationalities of overall development. It also raises the question of control over resources and the mode of their appropriation, all of which can radically alter the calculus of needs versus availability. Demand and supply of a resource are thus not uniquely determined, and the imbalance between them has to be contextually interpreted.

Further he goes on to review the various ecological side-effects of large dams (well known by now) which raises the question of decision-criteria for investment in large dams. In this connection problems of appraisal of projects are noted such as quality and criteria of approval as well as their subsequent monitoring. Predominance of an engineering perspective on these issues to the relative exclusion of social relevance of the projects is noted. Although prospective benefits from projects are typically to be weighed against costs as criteria of decision-making, problems arising from valuation of costs and benefits, and failure of benefits to materialise are also pointed out, problems related to implementation of projects, development and utilisation of irrigation potential created are also discussed. An overall awareness of the magnitude of the problem is thus reflected. But the remedial attempts suggested however do not reflect a corresponding perception of the working of the state machinery. The usual palliatives are offered, as for instance, with reference to the rehabilitation issue he says "the only answer to this is that rehabilitation measures should be *inspired* by the aim of not merely mitigating hardship but also assuring to the affected people a quality of life to ensure equitable distribution of benefits through *legislation* enabling attachment of land in command area to project affected persons" (Iyer, 1989 p A-11) (emphasis added). For effective monitoring he recommends a reviewing body "armed with ....powers of enforcement" (p A-12). For improving the quality of appraisal of major irrigation works the inculcation of a body of high level 'experts' from various disciplines (including agriculture, environment, economics, etc.) is suggested. Concerted planning, greater discipline, imposing stringent conditions for project clearance, building right attitudes through "extensive dissemination of information and analysis" and so on are some of the other prescribed efforts to ensure the viability of large dams on social, economic and technical counts. While sounding a note of caution regarding the workability of these measures as well Iyer winds up the discussion by positing it as the only way possible for resolving the problems since large dams will be taken up any way ! Thus, despite his insights regarding the problematic nature of large dams as they are instituted of date he prefers to submit to the strategy, rather than question it. It is this inference from his observations that constitutes his withdrawal from what could have been a more radical line of enquiry.

However, coming back to the point raised at the beginning, the basic issue with reference to irrigation is not whether major, medium or minor projects are proposed as appropriate. But that these considerations are made in the context of a development strategy which typically remains divorced from the needs of the people. It is the lack of people's involvement in the total process of conceptualisation, initiation, formulation and implementation and utilisation that constitutes the essence of the critique of large dams, as of the entire development strategy that is geared towards supplying developmental inputs from beyond (i.e., alien to) the social unit it proposes to benefit. It is this alienation from the people concerned that creates the disjuncture between what is provided and what is needed resulting in multifarious problems such as those observed in case of large dams. The technologistic perception of these problems in the literature and the resolutions offered for instance by invoking the assistance of experts, fails to bridge the gap between the people and the developers.

People's participation however is not merely their consultation at the level of decision-making. Rather it is to be viewed as an ongoing involvement at every stage starting from original conceptualisation of a particular scheme right through to its implementation and utilisation. The alternative trajectory of development being counter-posed is that which would evolve "from below" with the people in command. The critical difference here is people's power to veto the suggestions of the experts in a situation of clash of positions. The point is not to argue that problems will not arise given the alternative approach but that the choices made and resolutions thrown up will, in all likelihood, be more pertinent and relevant to the needs and problems of the people. Their involvement would constitute an insurance against manmade disasters. Our stance is by no means an anti-technology one as such but one where technological inputs are complementary to the efforts of the people and subordinated to them. Large dams as they are instituted at present for example, invert the order of hierarchy.

A further point to note is that the 'people' being talked of comprise distinct socio-economic categories with often conflicting interests. Developmental inputs such as large dams which have differential implications for different interest groups, invoke differing responses from various social groups. The question of technological inputs for development is therefore a political question. Attempts to 'technicalise' an issue that is basically social and political is therefore unwarranted as it happens for instance in reducing the problem with large dams to questions of their height! It is precisely on account of its political implications that people's participation does not find favour with the establishment. The control over decisions and resources, in the hands of people, could pose a serious challenge to status quo and the established power structure. Ironically enough even the organised left may not patronise such an approach precisely for the threat it may pose to the legitimacy of their 'organisation' as well. The alternative trajectory evolving from the grassroots therefore will have to struggle in the face of tremendous opposition by vested interests and can by no means be purely policy-induced.

The discussion on large dams also raises the much broader issue of the role of science and technology based inputs in the developmental process. The issue is not one of being pro or anti science. Neither can be maintained. But a strong case for questioning the rationality of modern science beyond the use/abuse model exists. If nothing else at least on grounds of historical experience with both -- the integration of modern science in the development strategy on the one hand, and pre-modern modes of appropriating nature by communities on the other hand. Neither can be a model for future alternative that needs to be explored--the historical clock cannot be pushed backwards. But it is imperative to recognise the following: one, the need to question science and technology inputs at a paradigmatic level; and two, to study the pre-modern systems of science and technology not out of romanticism, traditional or revivalist concerns but out of respect for their viability (in their historical context) born out of generations of experience in the material reproduction of societies, a point cogently argued by Sengupta (1985) with reference to irrigation.

To posit the issue as pro and anti science is therefore neither relevant nor tenable. However, it does go to the credit of the 'alternative wallas' to at least recognise and draw attention to the crises with the issues of development based on modern science and technology (S and T). The organised left on the other hand is content to fight the battles within the mainstream S and T paradigm or take refuge in attacking the revivalist tendencies among the alternative camp. While that may well be legitimate, this onslaught on romantic revivalism needs perforce to be combined with a fundamental critique of tenets of modern S and T. Vacillation on that score runs the risk of supporting the establishment science by default. Besides it blunts the *need* to search for an alternative which is the critical question today.

REFERENCES

Iyer, R., 1989, 'Large Dams: The Right Perspective', *Economic and Political Weekly*, Vol XXIV, No. 39, September 30, pp 107-115.

Sengupta, N, 1985, 'Irrigation: Traditional vs Modern', *Economic and Political Weekly*, Vol 20, Nos 45-47, pp 1919-1938.

THE LARGE DAM DEBATE : A RESPONSE TO AN INTERVENTION

Ramaswamy R. Iyer

I am very gratified at the careful and detailed attention that Jasveen Jairath has devoted (*EPW* January 13, 1990) to my article on the large dam controversy. Such a serious critique deserves a considered response, and I shall attempt it in the hope that others will join this debate.

Before proceeding to a discussion of Jairath's main thesis, may I question a couple of her incidental comments? Firstly, she seems to find the opponents of dams a more enlightened group and more aware of important issues than the supporters of such projects. I have been following the controversy with some diligence, and have to say with regret that partisanship, shoddy arguments, and a lack of awareness of larger issues, are to be found on both sides of the divide. One of the purposes of my article was precisely to make those larger implications explicit.

Secondly, Jairath has been quite complimentary about my article and it may seem churlish to cavil at praise; but may I say that I am somewhat chagrined at finding myself classified as an enlightened member of a particular set, namely, the supporters of dams? My whole effort was to stand aside from both sides to the dispute and arrive at the truth through patient questioning. I did not set out as a supporter of large dams. When I began writing my article the conclusion that I expected to arrive at was that while one could not say 'no' to such projects solely on environmental grounds, one might indeed have to say 'no' on a whole range of practical considerations; but in the process of writing the article I found that such a position could not be maintained. It was only gradually that I came to the conclusion that one had to be prepared to support a few such projects, though with extreme caution.

That brings me to Jairath's fundamental criticism of my article, namely, that I timidly withdraw from the implications of my arguments and prefer to submit to the strategy rather than question it; that I fail to pursue what could have been a more radical line of enquiry. I appear to have been misunderstood. My point was that within the universe of discourse set by the developmental pattern chosen by us, we could not isolate large dams for attack; and that on the other hand, if we were willing to challenge the whole course of development, there would no longer be any point in selecting dams for attack. I had in fact explicity stated that I was avoiding a discussion concerning developmental choices. However, for the purpose of the present debate I am willing to abandon that self-imposed limitation and enter into larger considerations. Having quoted a sentence from my article with approval, Jairath observes: "But that is precisely the point -- why not go forward"? Why not indeed. Let us following Jairath's reasoning and see where it takes us.

She finds the entire process of development "anti-people" and proposes two remedies: (i) people's participation ("at every stage from the original conceptualisation of a particular scheme right through to its implementation and utilisation"); and (ii) a re-examination of the role of science and technology in the developmental process, so that we move to an order in which technological inputs are complementary to the efforts of the people and subordinated to them (unlike the present situation which inverts that hierarchy). Alas, "people" are not necessarily enlightened. It is easy to whip up popular sentiments in favour of large projects (whether industrial plants or irrigation or hydro-electric projects), and opposition to such projects can be made to appear to be opposition to the interests of a particular state or region or locality or community and therefore to be anti-people. One could be in total agreement that people's participation is desirable, but remain unconvinced that it is a guarantee of wisdom or that it will necessarily lead to a change in the role of technological inputs in the process of development. The latter is indeed the fundamental issue, but Jairath is careful to say that she is not "anti-technology" or "anti-science" and that "the historical clock cannot be pushed backwards"; but why are these disclaimers considered necessary? If a radical change in the development process is indeed called for, why not accept the logic and advocate it? This part of Jairath's critique seems to me to be a mere adumbration -- an incomplete postulation -- of a radical argument. To address her admonition to herself: why not go forward? Let me at any rate venture into dangerous territory.

In my article I had described the developmental process as "demand-based, demand-multiplying, technology-driven and growth-oriented". The three essential components in this are: an exaltation of demand, an exaltation of technology and the acceptance of growth as an article of faith. The first leads to genuflection before the gods of production and consumption; this in turn means that we are firmly and inescapably mounted on the treadmill of growth; and the exaltation of technology means the failure to consider the possibility that the ever-accelerating pace of technological change could be a pathological, dehumanising condition rather than an indication of health. In so far as these crucial characteristics are concerned there is no difference between capitalist and socialist societies; and that is why the political and organisational differences between the two are now breaking down. The true – the human – alternative to this mode of development would be one which does not take it for granted that "civilisation consists in the multiplication and refinement of wants"; aims at stability, welfare and sanity rather than growth; and assigns centrality to the human being rather than to technology. The only consistent philosophy of this kind seems to be the Gandhian vision. (Jairath may say that that was not what she had in mind, but that is the direction in which her arguments about 'people' and 'technology' point.) That vision is profoundly attractive, but I am deeply troubled and uncertain about the tenability of such a choice.

I shall refrain from pursuing hypothetical speculations about whether at a certain stage in the past a different developmental choice could have been made, and consider instead the practical question of whether we can *now* turn away from the path already chosen to a different one. Where should we start? What large industries should we scrap? What industries should we retain and how should we prevent them from regenerating the inexorable growth process? If, as Baba Amte and others argue, we should say 'no' to dams above a certain size, should we not equally say 'no' to industries above a certain size and to industries of certain kinds ? It is easy enough to say that we should say 'no' to nuclear installations, but what about oil refineries and chemical industries ? They are all handlers of dangerous substances and operate under horrendous pressures and temperatures; each one of them is a potential Bhopal. Clearly, we should say 'no' to all of them; as we should to gigantic power stations, whether they burn coal or oil or gas. If we say 'no' to a large dam planned to be located in a seismic zone because of the potential danger of dam-burst and inundation, should we not say 'no' to the international traffic in oil, considering the statistical inevitability of accidents and the consequent damage from oil spills from tankers? However, under those conditions we can have no large industry, no coal-mining, no steel, no railways, not even large-scale commercial agriculture. Without energy and without large surpluses of foodgrains from the countryside, we cannot sustain Bombay, Calcutta, Madras, Delhi or Bangalore. An alternative vision must, therefore, necessarily mean a tremendous shrinkage of these cities. We may dislike these monstrous conglomerations but they exist and consist of human beings; their shrinkage will be a very painful process indeed.

Moreover, assuming that we make a choice in favour of the Gandhian alternative, would be able to pursue our own path undisturbed ? Would the world not inevitably break in ? The products and ways of living of the rest of the world would surely cast a spell on the minds of people in this country and make them discontented with their own civilisation, which, superior though it might be, might seem akin to poverty in comparison with the tawdry world outside ? We have repeatedly seen the serenity and good nature of "primitive" communities shattered by the onslaught of the modern world. At a higher level we have seen that designer clothes and disco dancing are beginning to cast a spell even on Chinese youth, not to mention the Russians who were beguiled much earlier by the blandishments of the western world. The tremendous turbulence in Eastern Europe has been caused not merely by the quest of political freedom and civil rights but also by the desire for consumer goods and consumer durables of the kind so evidently abundant in Western Europe. What grounds are there for supposing that a choice in favour of the Gandhian model of development would remain inviolate over the decades ?

The purpose of these questions is not rhetorical nor are they negatively intended. It is not being argued that the Gandhian alternative is impossible. The perplexities that are being posed (in the hope of eliciting answers) are whether the full implications of such a choice are clear to many who talk about it glibly; whether they would be willing to pay the price that it implies and accept the enormous hardships of a transition to that kind of world; whether, having made that transition, the community would remain (or be allowed to remain) stable and unaffected by the rest of the world; and whether indeed the Gandhian choice is possible for any one country unless the entire world is willing to make such a choice.

Some may feel that I am posing too big an Either/Or; that one does not have to make such drastic choices, but can make small decisions, one at a time. I wish I could believe this. Those who subscribe to Schumacherian ideas of "small is beautiful" or "intermediate (or appropriate) technology" will have to show how precisely they can be brought into operation. Once we accept the modern world and its technology, it seems difficult to stop at intermediate positions. Technology has a momentum of its own; each technological improvement or innovation may seem small but cumulatively they transform the world; and then the process of technological change itself begins to accelerate uncontrollably. That is our present position.

I hope that Jairath who found my earlier argument timid will not find the present one rash. If I have presented the matter in an excessively dramatic or apocalyptic manner, I am prepared to be corrected. I would respectfully submit to earnest environmentalists and humanists like Baba Amte that instead of attacking a particular manifestation of a system of development, they should challenge the system itself, not in negative terms of condemnation, but in positive terms by offering a viable alternative system with a detailed operational plan.

I propose to leave the debate at that indeterminate stage; I offer no certainties but merely raise questions. Meanwhile I have to revert to the point that in the world we live in, and pending a fundamental developmental choice, we cannot rule out large dams, but must be extremely cautious about undertaking them and surround them with safeguards, which will necessarily have to be (*pace* Jairath) a mix of technological, methodological, administrative and institutional measures and mechanisms, including but not confined to 'people's participation'.

6

Role of Major Dams In Indian Economy

R. B. Shah

INTRODUCTION

Right after the country achieved independence, one of the major economic activities undertaken was the construction of major dams for providing irrigation, hydro-electric power, flood control etc. After the Plan era commenced in 1951 this activity got a further impetus and almost through all the successive Plans irrigation and power occupied a prominent place of priority. The first Prime Minister of India, late Pt. Jawaharlal Nehru, realising the great potential of the major dams in the growth of Indian economy aptly called these dams as "Modern Temples of India". In fact, it was he with his vision and long range planning provided great inspiration and motivation to what was then called as a programme of construction of river valley projects. The most important and significant projects taken up immediately after independence were the construction of Bhakra Nangal Project on the Sutlej River in the North, a series of four dams on the Damodar River in the East along with Hirakud Dam on the Mahanadi River and the Nagarjunasagar on the Krishna in the South. By any standard, these projects were really massive particularly in the context of the nascent economy of the country then. These projects even today continue to be some of the largest and the highest in the world. For example, Hirakud Dam is still the longest earth dam in the world, Nagarjunasagar the largest masonry dam, and the Bhakra Nangal, one of the highest concrete gravity dams. It is an indisputable fact that if there was any single prime mover the prosperity in the states of Punjab and Haryana, it was the Bhakra-Nangal Project. With the irrigation and the power available from this project, these two states not only became the top rankers among the Indian states in the economic ranking-both on the agricultural and industrial front-but have ever since contributed a lion's share to the overall food production in the country.

Even these few facts bear an eloquent testimony to the role of major dams in the Indian economy. However, of late, unfortunately, there is a regular and sustained campaign of dis-information regarding the role of major dams, ignoring some of the basic facts about the country's land and water resources.

BASIC INDISPUTABLE FACTS

Land Resources

Let us therefore have a broad overview of some of the basic indisputable facts regarding the country's land and water resources. The geographical area of the country is 328.8 million ha, of which the culturable (cultivable) area is 186 million ha which is 56.5 % of the geographical area. The net sown area in 1985 was 141 million ha which is 76 per cent of the culturable area. It is also well-recognised that there is no possibility of extension of culturable area on the other hand, it is likely to decrease on account of increasing pressure of land for urbanisation and other developmental works.

Water Resources

The basic and the most important input for agricultural production as well as sustenance of human and animal life is the water. In India, the predominant source of water is rainfall which, on an average, is about 400 million ha per annum. However, as is well known, the rainfall in India exhibits great variations from year to year as well as in a year from month to month as well as in the areas distribution. The temporal and spatial variations in the occurrence of rainfall result in drought-flood-drought syndrome. About one-third of the country is subject to droughts and about one-eight is subject to floods. Both floods and droughts cause extensive damage almost every year and thus adversely affect the national economy. The impact of the recent droughts on the Indian economy is too well known to be forgotten soon.

The available surface water resources of the country have been assessed at 180 million ha m. However, since about 80 % of this water flows during 3-4 months of the monsoon, this cannot be utilised unless stored in reservoirs for subsequent use. On account of limitations of topography, we do not have enough dam sites to store a major portion of the available water. It has, therefore, been estimated that only about 69 million ha m of surface water can be utilised which constitutes only 38 per cent of the available water resources. In addition, the country has about 42 million ha m of utilisable ground water. In 1951, the actual utilisation of surface water was about 20 per cent of the utilisable quantum and 10 per cent of ground water. It is estimated that, as of now, the country is utilising about 50-55 per cent of the utilisable quantum of surface waters and 50 per cent of ground water.

NEED FOR STORAGE DAMS

Till a few years back, there was a feeling in the country that we have abundant water resources. However, it is only of late that it is being increasingly realised that water is a precious and a scarce natural resource even in India and demands on the water for various purposes are fast increasing. From the facts stated earlier, it is obvious that the country can ill-afford the wastage of available water flowing into the Seas and causing damage by floods en route. It is absolutely necessary to conserve as much water as possible by small and big storage dams for use throughout the year. It can also be easily appreciated that since the quantum of water to be stored is considerable, it would be possible only through major dams, as the total quantum of water to be stored through minor projects will be a very small percentage, as well be evident from the figures given subsequently.

Drinking Water Supply

Although irrigation is and will continue to be the bulk consumer of water even in the future, the other uses of water also call for construction of storage dams. I will first deal with these aspects before coming to the subject of irrigation. The requirements of drinking and municipal water supply to metropolitan and other important towns in the country have already become critical. For example, the entire water supply of Bombay is dependent on a series of dams like Vaitarana, Tansa and Bhatsa. In fact, there is now a proposal to construct another dam on the Vaitarana to meet the increasing needs of Bombay's water supply. Water of Pune town in Maharashtra is met with from Panset and Khadawasla dams. Hyderbad is mainly dependent on its water supply from Manjira and Singur Dams. Delhi, which is situated on the banks of the Yamuna gets hardly 25 per cent of its needs from Yamuna. The balance needs are met with from releases from Bhakra Dam on the west, and Ram-ganga Dam on the east. For the future needs of Delhi, it is proposed to obtain water from the Tehri Dam being constructed in the north on the Ganga. This is in addition to a number of tubewells, which contribute hardly 10 per cent of Delhi's water supply. The acute scarcity of water supply in Madras is too well-known. Since apart from the storage available in Poondi Reservoir, there was no other possibility of further augmenting the water supply, the States of Andhra Pradesh, Maharashtra and Karnataka agreed to part with their allocated share of Krishna waters to provide 15 TMC of water supply to Madras city which will be made available through the stored waters at Srisailam Reservoir in Andhra Pradesh carried through 430 km long canal to reach Madras city. A major dam at Bisalpur is being now constructed to provide water supply to Ajmer city. Warrangal town in Andhra Pradesh depends for water supply on Sriramsagar Dam.

Industrial Water Supply

Although the industrial requirements of water constitute only a small per cent of the total consumption of water, if again cannot be met with without the construction of storage dams. For example, for the needs of Bokaro Steel Plant, another dam at Tenughat in the Damodar River Basin had to be constructed. Similarly, a cluster of thermal and superthermal stations in U.P. are entirely dependent on the waters stored at Rihand Dam on a tributary of the Sone River in U.P. Similarly, for the Vishakapatnam Steel Plant, the water is to be supplied from Yeleru Reservoir under construction in Andhra Pradesh.

Floor Control

As already mentioned earlier, one-third of the area of the country is subject to recurrent floods. One of the surest methods to control floods is to store the excess waters in our major dams. The Hirakud Dam built on the Mahanadi has considerably mitigated the problem of floods in the rich and fertile Mahanadi delta. Similarly, the series of dams constructed in the Damodar Basin with one of the principal objectives of flood control have again substantially controlled the recurring flood menace in the Damodar Basin. The Bhakra Dam itself has completely solved the problem of recurring floods in the Sutlej river, where a large extent of the bed of the Sutlej below the Bhakra Dam is now utilised for intensive agricultural use. Teh Surat town, one of the richest in Gujarat, which used to suffer serious damage on account of floods is now completely protected by the Ukai Dam. Similarly, the problem of floods in quite a few rivers has been mitigated by construction of dams although not specifically designed for flood control.

However, the problem of floods continues in the Ganga and Brahmaputra Basins, as no dam has so far been constructed in the Brahmaputra Basin and there is very small scope for construction of dams on the major tributaries of the Ganga in the states of U.P., Bihar and West Bengal as suitable sites for construction are not available within our territory.

Irrigation

As far as irrigation is concerned, out of the total irrigated area of 22.6 million ha in 1951, 9.7 million ha was the contribution of major and medium projects. The irrigated area at the end of the Sixth Plan was 68 million ha out of which 30 million ha was on account of major and medium irrigation, mainly through construction of a number of major dams.

POSITION OF STORAGES IN INDIA

Before I deal with the position of storages in India, it is significant to note that the available water resources in India and in the United State of America are almost identical. However, the storage capacity already created in the United States is about six times that of India.

Before independence, there were thousands of small tanks constructed in the States of Andhra Pradesh, Tamil Nadu, Rajasthan, etc. There were very few major dams like Mettur, Krishnarajasagar, etc. After Independence, till now more than 2,000 major dams (more than 15 m in height) have been constructed. The dams constructed so far have a live storage of 14.3 million ha m, which is about 50 per cent of the likely total storage capacity possible in India. The dams under construction have a live storage of about 8 million ha m and the projects contemplated in future have also a live storage of about 8 million ha m. Thus, the total storage, capacity likely to be created by major dams in India is about 30 million ha m, which is about 40 per cent of the utilisable surface water resources and only about 16 per cent of the available annual surface water resources. The total storage capacity of the thousands of tanks constructed aggregates only 3 million ha m or only 10 per cent of the storage capacity through major dams.

Even without an expert knowledge, it can be appreciated by any intelligent person that given the situation of the rainfall and consequently the pattern of river follows in our country, most of the waters cannot be utilised unless a major portion is stored during the monsoon for subsequent use for various beneficial purposes. It is also clear that the substantial portions of the waters can be stored only by major dams. One example of the Sone River will clearly bring out the effect of storages on the utilisation of water. The whole of Sone Basin traversing the states of Madhya Pradesh, Uttar Pradesh and Bihar is drought-prone and agriculture is the only foundation for the economic development in the basin. The surface water resources of the Sone on a 75 per cent dependable basis are estimated as 14.25 million ac ft. Before the construction of the Rihand Dam, the annual utilisation from diversion was only 20 to 25 per cent. After the construction of the Rihand Dam, the utilisation of waters almost doubled to 5.75 million ac ft. At present, another major dam on the Sone is under construction at Bansagar. With the completion of this Dam, the total utilisation will increase to 8.75 million ac ft, i.e. about three times the utilisation without storages. There is no possibility of constructing additional major dams of the type of Rihand and Bansagar. However, there are five other medium dams which are under construction on the tributaries of the Sone. After completion of these, the total utilisation will increase to 11 million ac ft, which will be able 75 per cent of the 75 per cent dependable flows. Even so, about 25 per cent of the waters of the Sone will continue to flow waste. Similar situation exists in most of our river basins.

IMPACT OF IRRIGATION ON AGRICULTURE

India has been and continue to be a predominantly agricultural country, where, even now, almost 70 per cent of the population subsists on agriculture. Consequently, the agricultural production plays a very vital role in the National economy. One of the basic and the most important inputs for agricultural production is water. Even green revolution would not have been possible without the basic input of water.

Means of Raising Agricultural Production

Agricultural production is raised by the following ways:

(i) Extensive cultivation, i.e. by extending cultivation to more areas of land.
(ii) Intensive cultivation, i.e. by raising more than one crop in the same area, and
(iii) Increasing productivity.

Introduction of irrigation brings more areas of uncultivated land under cultivation. Besides insuring Kharif crops against failure of monsoon, it enables growing Rabi crops in areas where it is not being raised. Irrigation also is the mainstay in increasing productivity as the high-yielding varieties of crops and increased use of fertilisers cannot sustain without assured water supplies at critical times. Thus, in all three means of advancing agricultural production, irrigation plays the key role.

Increase in Net Sown Area

Table 1 and Fig. 6.1 show the picture of land use and development of irrigation and agricultural production during the period 1951 to 1985, i.e., from the First Plan to the end of Sixth Plan. It will be observed that there has been very little increase in the net area sown since mid sixties and it is more or less stabillised at 141 million ha. The National Commission of Agriculture, in their report in 1972, had projected that the net sown area in 1985 will be 145 million ha and in the year 2000, 150 million ha. As against this, in 1985 we are still at 141 million ha and it is, therefore, doubtful whether the sown area will increase to 150 million ha by 2000 A.D. The options, therefore, narrow down to intensive cultivation and increased productivity.

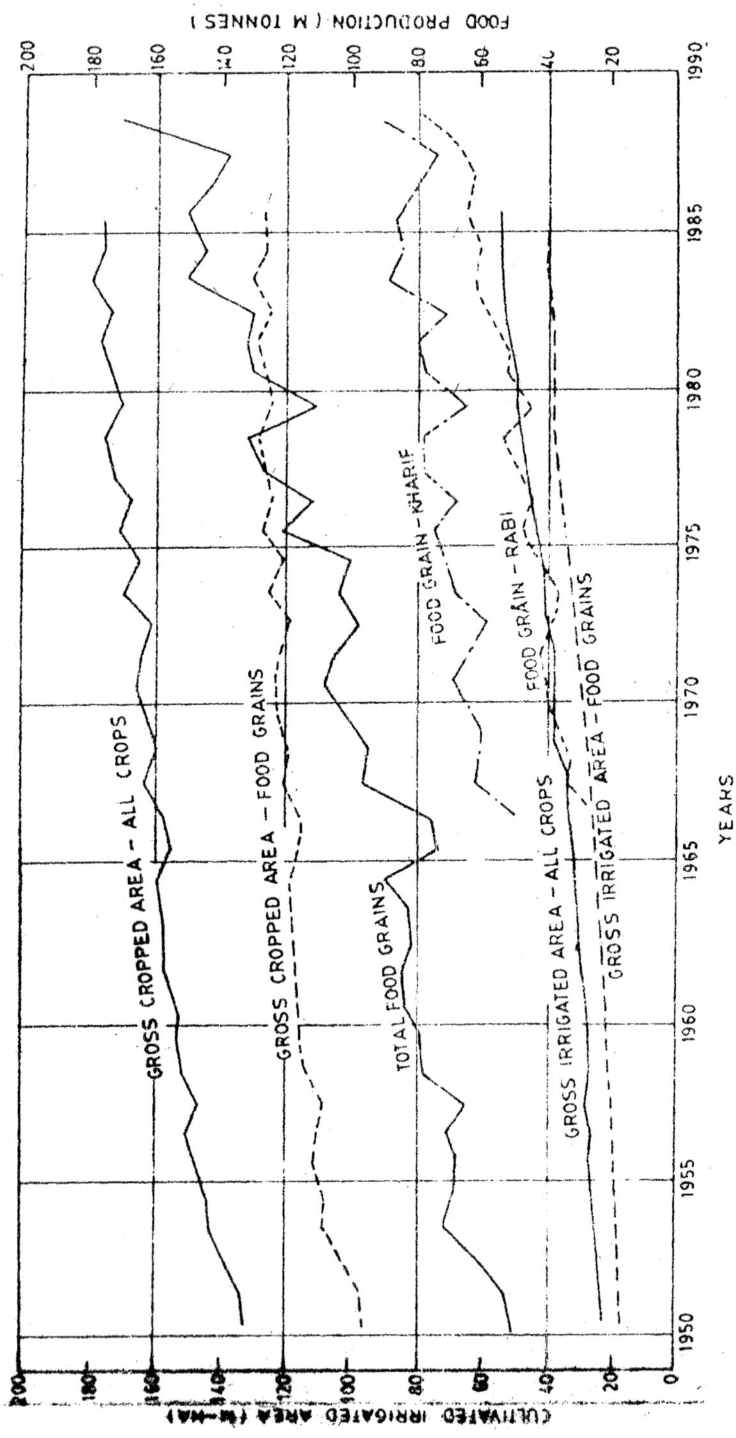

FIG. CULTIVATED AREA, IRRIGATED AREA AND FOOD PRODUCTION FROM 1950 TO 1989

Table 6.1 : Cultivated Area, Irrigated Area and Food Production From 1950 to 1985 Area in million ha and production milli on tones

	Area Sown			*Irrigated area*			*Food Grain Production*				
	Net	*Grcss Food grain*	*All crops*	*Areas Sown more than once*	*Net*	*Food grain*	*Grosss Area All crops*	*Irrigated more than once*	*Kharif*	*Rabi*	*Total*
1950-51	118.8	97.3	131.9	13.1	20.9	18.3	22.6	1.7	--	--	50.8
1960-61	133.2	115.6	152.8	19.6	24.7	22.1	28.0	3.3	--	--	82.0
1965-66	136.2	115.1	155.3	19.1	26.3	--	30.9	4.6	--	--	72.3
1966-67	137.2	115.3	157.4	20.1	26.9	--	32.7	5.8	48.9	25.3	74.2
1967-68	139.9	121.4	163.7	23.9	27.2	--	33.2	6.0	60.8	34.3	95.1
1968-69	137.3	120.4	159.5	22.2	29.0	--	35.5	6.5	59.6	34.4	94.0
1969-70	138.8	123.6	162.3	23.5	30.2	--	37.0	6.8	62.4	37.1	99.1
1970-71	140.3	124.3	165.8	25.5	31.1	30.1	38.4	6.9	63.0	42.2	108.2
1971-72	139.7	122.6	165.2	25.5	30.2	--	37.0	7.1	68.9	39.5	108.4
1972-73	137.1	119.3	162.2	25.0	31.8	--	39.1	7.2	58.6	38.5	104.7
1973-74	142.4	126.5	169.9	27.5	32.6	--	40.3	7.7	67. 8	36.9	104.7
1974-75	137.8	121.1	164.2	26.4	33.7	--	41.7	8.0	59.1	40.7	99.8
1975-76	141.7	128.2	171.3	29.6	34.6	34.1	43.4	8.8	73.9	47.1	121.0
1976-77	139.5	124.4	167.3	27.9	35.2	--	43.6	8.4	66.5	44.7	111.2
1977-78	142.0	127.5	172.2	30.3	36.6	--	46.0	9.5	77.7	48.7	126.4
1978-79	143.0	129.0	174.8	31.8	38.1	37.2	48.3	10.3	78.1	53.8	131.9
1979-80	138.9	125.2	169.6	30.7	38.5	37.8	49.2	10.7	63.3	46.4	109.7
1980-81	140.3	126.7	173.1	32.8	38.8	37.6	49.9	11.7	77.6	52.0	129.6
1981-82	142.0	129.1	177.1	35.0	39.8	38.2	51.6	11.8	79.4	53.9	133.3
1982-83	140.9	125.1	173.4	32.6	40.7	38.4	52.1	11.4	69.4	59.6	129.5
1983-84	142.8	131.2	180.2	37.4	42.0	40.2	53.9	12.0	89.2	63.1	152.4
1984-85	140.9	126.7	176.4	35.5	41.7	40.1	54.1	12.4	84.5	61.0	146.5
1985-86	141.1	127.1	177.3	36.3	41.8	39.9	54.0	12.0	86.0	64.5	150.5
1986-87	--	--	--	--	--	--	--	--	--	--	143.4
1987-88	--	--	--	--	--	--	--	--	--	--	138.4
1988-89 (anticipated)	--	--	--	--	--	--	--	--	--	--	170.0

Source : "Indian Agriculture in Brief" Area, Production and Yield of Principal Crops" published by Diirectorate of Economics and Statistics.

Increase in Gross Irrigated Area

The gross irrigated area has increased by about 31.4 million ha during the period 1951 to 1985. This has contributed to the increase in area sown more than once which has been 23.2 million ha and also to the provision of assured water supply to Kharif crop in times of drought or delayed rainfall, it is worth noting that the increase in the area under food grain crops during 1951-1985 has been about 29.8 million ha and that under sugar-cane about 1.2 million ha during the same period. This again is attributable to irrigation.

Increase in Food Grain Production

In 1951, the food grain production was about 52 million tonnes. In the current year, a record production of over 170 million tonnes is expected. Thus, there will be an increase of about 120 million tonnes in the period 1951-1989. This has been possible mainly on account of the irrigation. However, there is no denying the fact that while irrigation is one of the basic inputs for increasing productivity, in order to have optimal levels of production, other inputs like proper quality of seeds, use of high-yielding varieties of crops, optimal use of fertilisers have also to be ensured. It has been proved in various National demonstration farms in actual field that given other adequate inputs, irrigated areas can give an yield of 3 to 4 tonnes per ha. Although these levels of productivity have not yet been achieved in all irrigated areas, in Punjab and Haryana states as a whole, the yields of rice have been 3.5 tonnes per ha while the yields of wheat have been 3.6 to 3.8 tonnes per ha in irrigated areas as per statistics published by the Union Ministry of Agriculture. Similarly, in Tanjore District of Tamil Nadu in the Cauvery Delta, about the yields of rice have been as high as 5 tonnes per ha. With the present forecasts the country's population by 2000 A.D. will be about 1000 million, which will need a food production of about 270 million tonnes, which will further rise to 365 million tonnes by 2025 A.D. Even with the development of full irrigation potential of 113 million ha, about 40 per cent of the country's net sown area will be dependent on the vagaries of rainfall.

Impact of Major and Medium Irrigation

Fig. 6.1 clearly brings out that there is a rise in Rabi production in drought years where even the Kharif has declined. The share of Rabi production in the total food grain production is showing a steady rise to a level of about 46 per cent in 1987-88. It is obvious that irrigation, and specially major and medium, is responsible for this phenomenon.

The experience of three consecutive drought years has indicated that in the drought years, minor surface irrigation tanks dry up soon and the ground water levels deplete fast. In the worst drought of 1987-88, major and medium irrigation was the mainstay of food production. For example, it was possible to provide irrigation facilities to 2.3 million ha out of a total of 2.9 million ha of the areas under the comand of major and medium irrigation projects. As against this, it was possible to provide irrigation facilities to only 0.06 million ha out of a total command of 1.7 million ha from the minor irrigation projects. Although similar statistics in respect of other States are not available, more or less the same situation prevailed there also. This underlines the importance of major dams in combating the drought years in sustaining food production. This is also proved by the fact that in spite of the worst drought, the total agricultural production was 138 million tonnes, compared to an earlier peak production of 155 million tonnes achieved in 1983-84, giving a shortfall of only 11 per cent from the peak output.

Similarly, the likely agricultural production in the year 1988-89, which is expected to be around 170 million tonnes, the highest ever achieved in the country so far, proves the importance of irrigation in achieving peak-production in good years of rainfall, which, in turn, can help tide over the situation in years of inadequate rainfall, as in the previous three years.

DAMS AND ECONOMIC DEVELOPMENT

Hydro Power

The role of power in the national economy is too well known to need any elaboration here. In fact, the per capita consumption of power is one of the indices of national prosperity. As per the long-term National Power Plan formulated by Central Electricity Authority, the country will need an installed capacity of about 174600 MW by 2000 A.D., which means that additional growth of about 10 per cent every year during the entire period will be necessary. The hydro-power potential of the country is estimated at about 442 million units of firm annual energy and 84044 MW of firm power at 60 per cent load factor. Unfortunately, so far, the installed capacity in hydro-power today is only 17650 MW, which utilises 13.6 per cent of the ultimate potential. With the completion of on-going and sanctioned schemes, the exploitation of hydro-power potential will still be only 20 per cent. This underlines the importance of speedier exploitation of hydro-power in future which is one of the Nature's gifts to the country. It is also well known that hydro power is totally pollution-free. Unlike thermal power and it uses the renewable resource, viz. water, unlike the progressive diminishing of coal reserves by use of thermal power. Besides, the cost of general of hydro-power is less compared to thermal and nuclear power generation, besides meeting the needs of peak demands which can be done only through hydro-power.

It is sometimes aruges that the hydro-power can be exploited by a series of small dams or diversion structures rather than constructing major dams. The futility of this argument can be seen from two facts. Firstly, the quantum of hydro-power which can be generated by small dams naturally will be very limited, particularly in the context of the ultimate power scenario. Secondly, detailed studies have indicated that the cost of one kW of installed capacity increases three-fold in a small dam compared to a large dam. Another example will clearly bring out the role of major dams. In the Chenab Basin, a number of hydro schemes have been planned through river diversions. It is also proposed to construct one major dam at Bursar in the upper reach. With that single storage and the regulated releases therefrom, the hydro-power from a series of downstream diversion projects gets doubled from 10,000 MW to 20,000 MW.

Economic Development Through Irrigation

Direct Benefits : Detailed studies of the gross income, cost estimates and net income per ha of irrigated areas indicate that the irrigated areas would result in an increase in net income of about Rs. 3,000 to Rs. 4,000 per ha at 1982-83 prices. A study of the Rajasthan Canal Project, Phase I, carried out by the National Council of Applied Economic Research some years back indicate that gross receipts per man-day in irrigated areas have increased by more than 100 per cent in all economic activities. Irrigated farming also results in increase in permanent employment which is of the order of 30 to 50 man-days per ha, or about 60 to 100 million man-days per year at present growth rate. Studies also indicate that the manpower used per ha of cropped area is about three times higher in irrigated agriculture as compared to un-irrigated agriculture. The obvious reason is the elimination of risk in crop cultivation due to availability of irrigation facilities resulting in higher yields and consequently more returns from agriculture.

Indirect Benefits The overall economic development of Punjab and Haryana in the last two to three decades reveals that there exists a close relationship between water resources development and structure of the economy. With the availability of adequate water and power, and with the adoption of new agricultural technology, the transmission of growth impulses from one sector to another can take place in several ways. The inter-dependence of production activities arises from the fact that each production activity demands input supplied from other production activities. A sector is linked with other sector which supply inputs to it and also with those which use outputs as their own inputs.

The linkages induced by input demand and those induced by output supply thus, generate growth impulses that are transmitted from one sector to the others. As regards links to manufacturing industry, a shift to improved technology creates demand for inputs like chemical fertilisers, pesticides, irrigation pumps and motors, improved implements and other agricultural machinery. The growing demand for these inputs in turn lead to setting up of industrial units for manufacturing these inputs, as well as establishments for servicing and repairs of machinery and implements thereby generating additional employment opportunities in non-farm sector generating additional income. It also gives rise to agro-based industries as well as other small industries by attracting the surplus income generated. The development of capital intensive industry as well as increased surplus agricultural production also requires adequate and extensive transport system to move raw materials and finished goods, communications, housing, widespread and efficient wholesale and retail trade network and other many allied services and utilities. In fact, in early stages of development, services and utilities which provide the vital infrastructure for industrial development may grow even faster than industry in terms of both employment and income.

The increase in crop output provides adequate feeds and fodder for livestock and thus accelerate animal husbandry activity. Besides, the vast water spread promotes fisheries and forestry as well as to a greater level. All these would obviously result in additional income which would create a demand for additional production and distribution services to handle and process the increased output and deliver the final product to the consumer. In addition to the new employment generated in storage, transport, processing and distribution of greater output of the agricultural sector including crop and animal husbandry, fisheries and forestry, employment will be created in the construction of infrastructure facilities like feeder roads, storage godowns, manufacturing yards, etc., required to carry such activities. There will thus be various multiplier effects from the adoption of new agricultural technology and increase in cropping intensity made possible through the expansion of irrigation facilities.

The increase in personal disposable incomes also creates demand for services like transport, communications and recreation. The increasing preference for consumer durables like cars, scooters, refrigerators, radios, T.V.s and kitchen gadgets will also need the services catering to their repair and maintenance. The process of economic development while on the one hand will lead to large governmental activities in the form of regulation and administration, on the other hand it will bring about a rapid change in the life styles of people calling for more services in the field of housing, education, entertainment, health and tourism. Such an overall change in socio-economic scene has already been experienced in Punjab and Haryana on a very wide scale and in some of the other irrigation projects like Rajasthan Canal, etc. on a lesser scale.

Average receipts for household with irrigation is 2.2 times more than households without irrigation. In fact the receipts from crop farming alone are found to be almost 6 times higher for households with irrigation.

It is an established fact that as the income rises, relatively greater part of it is spent on productive activities and less on consumption. From the analysis of RCP it has been seen that the consumption expenditure in case of non-irrigated areas is 73 per cent of the total expenditure. On the other hand, in irrigated areas it is only 57 per cent balance accounting for productive activities. Further examination of consumption pattern also brought out that people with higher income spend relatively less on food and more on non-food items as compared to those with lower income.

The analysis further indicated that with increase in income, capacity to save and invest also increased. It is estimated that the investment made by household with irrigation was higher by 57 per cent than those without irrigation. It was also observed that those with irrigation invest in farm equipment in a big way for further increasing the income.

This clearly brings out the higher standard of living of the people in irrigated areas. It further proves that in irrigated areas the people have been able to satisfy their demand for food and they are using increased income in other non-food items. This itself sets in motion the forces of economic development.

CONTROVERSY BETWEEN MAJOR AND MINOR DAMS & GROUND WATER

An unnecessary controversy has been raised in recent times between the major dams and minor dams on the one hand and the major dams and use of ground water on the other hand. In fact, this controversy is totally uncalled for and unnecessary, as minor dams and use of ground water are not an alternative to major dams but are supplementary. Out of the ultimate irrigation potential of 113 million ha, ground water can provide only 40 million ha on the basis of the annual rechargeable ground water. Similarly minor dams can provide irrigation to only 15 million ha; while major and medium dams together can provide irrigation facilities to about 58 million ha. Construction of dams is basically a location specific issue. A dam, major, medium or minor, cannot be constructed at any place, as it needs certain typographical conditions to make it economically viable. For example, to provide drinking water supply to Delhi, there is no possibility of constructing a dam on the Yamuna anywhere in the vicinity of Delhi, excepting Kishau dam in the upper reaches of the Yamuna.

Relative Economics of Major and Minor Dams

To replace a single large dam irrigating 100,000 ha, 10 to 20 small dams each irrigating 5000-10000 ha would need to be constructed either in upper or lower reaches of a watershed. Such a large number of alternative sites is rarely available in practice necessitating a curtailment in the envisaged development, if the latter alternative is preferred. Construction of dams in upper reaches involves problems in transporting heavy construction machinery and paucity of local skilled labour. The submergence in upper reaches is usually valuable forest land. Construction of dams in lower reaches involves more submergence (of usually cultivated land) per unit of storage due to flat nature of topography.

The recent experience of Heran Dam in Gujarat State will not be out of place in the present context. The construction of this dam has not yet been taken up because of agitation of local population (that would have been displaced), supported by some environmentalists. Preliminary evaluation of an alternative proposal to reduce the height of the dam and construct a series of small dams upstream showed that the population affected and land submerged reduced only marginally but the submergence of forest land and tribal habitats increased tremendously. As no other alternative supply of irrigation water was found feasible, the only result has been denying irrigation facility to good, cultivated land which would have otherwise been a boon to the people. Other data available in CWC also show that when more storages are constructed to substitute a single large storage the cost and submergence are invariably more. To compare the cost and submergence aspects of large dams with that of small dams, a study was made for Jonk sub-basin in Mahanadi considering two alternatives. In alternative A, only Girna dam, a major project with storage of 603 Mm^3 was considered while alternative B considered substantial reduction of the storage at Girna to 185 Mm^3 and making up the rest 418 Mm^3 at 8 other identified storage sites. A rough estimate of cost of storage only showed that the alternative B was costlier than alternative A by 150 per cent and involved 60 per cent more submergence. The evaporation loss in alternative B would obviously be accordingly more, because of larger waterspread, by as much as 50 to 60 per cent.

Sometimes, the capital cost of minor irrigation tanks is brought out to indicate how they are cheaper than major dams. There is no doubt that the capital cost of a minor tank is much less. However, firstly these cannot replace the major dams. Secondly, these are normally not provided with any spillways to cater for safe disposal of flood waters. Consequently, breaches are very common involving higher maintenance cost apart from loss of crop and cattle, losses in distribution systems, etc. These aspects need to be taken into account in working out the cost tanks in Tamil Nadu, Karnataka and Rajasthan has indicated that in years of high rainfall, there are series of breaches in tanks like a chain action. The cost of major repairs to these tanks in the events of breaches also has to be taken into account.

Use of Ground Water

Ground water is not a distinct entity but is a part of the same quantum of water forming a part of the same quantum of water forming a part of a hydrological cycle. There is a two-way flow between surface and ground water. In fact, in most of the rivers, the low flows are contributed by the ground water. The ground water can therefore supplement and augment the use of surface water but it cannot replace major dams. In fact, an erroneous feeling of unlimited availability of ground water resulted in over exploitation in a number of States resulting in permanent damage. For example, in Saurashtra Region of Gujarat, over-exploitation resulted in increase of salinity in the ground water aquifers; consequently making the cultivated areas going out of production. The use of ground water is also highly dependent on the availability of power.

It is often advocated that the cost of ground water development is very cheap compared to the use of surface water through major and medium irrigation. This again is not true, as only the capital cost in both the cases is taken into account. As is well known, the recurring annual maintenance cost in the case of ground water is considerably high. Besides, normally the life of the tubewells is about 20 years, while the useful life of the pumps is only about ten years, needing complete replacement after the useful life. Therefore, if the annual costs over a period 100 years, which is taken as the economic life of a mjaor project, are compared taking into account both the capital and the annual maintenance and replacement costs, the picture that emerges is as given in Table 6.2.

It will thus be observed that even taking the present cost of major and medium irrigation, the annual cost becomes cheaper compared to the minor tanks and ground water. Even so, in order to meet the ultimate needs of the country for agricultural production, irrigation has to be provided through all the three means, viz. major and medium projects, minor surface projects as well as by ground water.

Table 6.2 Average Annual Cost of Major/Medium Irrigation and Minor irrigation

(Rupees per ha)

	Items	*Major & Medium*		*Minor Projects*	
		Storage Schemes	*Diversiion Schemes*	*Surface Schemes*	*Tubewells*
1.	Capital Cost	30,000	20,000	13,000	7,500
2.	Operation and maintenance cost per year (inclusive of establishment and energy)	300	300	500	665
3.	Expected life of projects in years	100	100	30	20
4.	Total operation and maintenance cost for the life of the projects	30,000	30,000	15,000	13,300
5.	Total cost	60,000	50,000	28,000	20,800
6.	Average annual cost	600	500	933	1,040

VALIDITY OF CRITICISM AGAINST MAJOR DAMS

In the last decade, there is a trend of mounting criticism and a deliberate campaign of dis-information about the so-called negative aspects of major dams.

Increasing Cost

It is often quoted that cost per ha of irrigation through major and medium dams has increased from about Rs. 1530 per ha in the First Plan to about Rs. 30,000 per ha in the Seventh Plan. It is, therefore, argued that investment of this order is not in the national interest and should be stopped forthwith. Firstly, the comparison of these figures is fallacious, as these do not take into account the element of inflation which has been the most predominant factor in the increase in cost, which is not applicable only to the irrigation sector but to every sector of the economy in the country. An exercise was, therefore, carried out to have a picture on the basis of constant prices with 1970-71 as the base. Adopting a Plan-wise deflator worked out by the Planning Commission, it is observed that the cost per ha through major and medium irrigation at constant prices (1970-71) works out as under :

First Plan to Fourth Plan	Rs. 3340-3940
Fourth Plan to Sixth Plan	Rs. 4300-4600
Sixth Plan	Rs. 5700
Seventh Plan	Rs. 5830

It will thus be observed that at constant prices the increase from the first three Plans to the Seventh Plan is about 60 per cent higher and not 2000 per cent higher as brought out by the critics. There are some valid reasons for the increase in cost, even at constant prices. Some of the main reasons are as follows :

(i) The projects taken up in the earlier period were, by and large, on better sites with lesser geo-technical problems.
(ii) There has been considerable increase in the cost of providing adequate rehabilitation measures in the Projects recently.
(iii) The distribution system in the earlier projects was terminated at 40 ha blocks, which is now carried up to 5-8 ha blocks.
(iv) On account of greater awareness for presentation of environment and ecology, additional expenditure is now incurred on compensatory afforestation, etc.

Submergence of Forest Areas

It is reported that the forest are in the country is shrinking at a rate of about 1.5 lakh per year on account of various factors. The major dams have been responsible for only 12 per cent of the forest land losses during the period 1951 to 1985. While considerable noise is being raised about this loss, not much is being said or done for the 8 per cent loss of forest area by other causes. Besides, one has to realise that construction of storage means submergence of some area. Obviously, in view of the limited cultivable area, the country can ill-afford the loss of that area. Besides, since most of the dams are located in the upper reaches of a river basin, the submergence has to be on waste land or forest land.

The submergence of forest land has also to be viewed in the contest of the benefits of irrigation over a large area. Analysis of data of a number of major projects in the country indicates that the total area submerged by construction of dams varies from 3 to 10 per cent of the area to be irrigated by these projects. The submergence of forest area is only about 1-2 per cent of the area irrigated. It is thus obvious that the loss of forest land is a very insignificant percentage of the economic benefits to a large area of land.

However, now, the forest land submerged by major dams is replaced by compensatory afforestation at project cost. Therefore, as of now and in future, there will not be any net loss of forests due to major dams.

Displacement of Population

This again is one of the major criticism against major dams. Any development activity involves some displacement. A similar study carried out in respect of a number of major dams indicates that the population displaced on account of construction of dams is about 2 to 4 per cent of the population benefitted by irrigation facilities which again is a very small percentage. It is a fact that in the past adequate care was not taken of the displaced population. However, this defect has now been rectified and it is ensured at the project cost that the displaced population is rehabilitate in such a fashion that they get better standard of living than in the existing pre-dam conditions.

Sometimes, it is also argued that the Tribal people should be displayed at all, as they cannot live in a different environment. This approach is again faulty and will result in perpetuation of their backwardness for times to come. In fact, the Tribal population has not yet joined the mainstream of economic development as no efforts have been made to improve their lot. Experience in the North-Eastern region indicates that the Tribal people definitely want water storage projects for bringing in overall economic prosperity to them.

Waterlogging

An opinion also prevails that large cultivated areas are affected by waterlogging as a result of introduction of irrigation through surface waters. Some highly exaggerated figures are quoted by some people. While it is a fact that some of the major irrigation projects experienced the problem of waterlogging, the area of land waterlogged is generally very small as compared to the irrigated area. The data available in the CWC indicates that the percentage of waterlogged area varies between 1.5 per cent to 10 per cent of the benefited area in various projects. However, in order to reduce even this extent, a number of measures are now being taken to water-logging by provision of adequate drainage network, increasing uses of ground water, etc.

Delayed Returns fror Major Projects

Although, out of 246 major irrigation projects taken up during the Plan period 1951-85, 181 are still ongoing in the Seventh Plan, it would not be correct to say that the investment on these has not yielded any benefit so far. Of these, there are only 57 projects belonging to the pre-Fifth Plan period (1951-74) and 63 per cent of the their ultimate potential has already been realised by 1985. Most of the remaining projects are in the initial stage, but even so have created about 8 per cent of their potential. Even so, this is not a happy state of affairs and from Seventh Plan, Government of India is not encouraging taking up of new projects by the states which in the past on account of socio-political factors took a number of projects resulting in thin spreading of resources.

Lag in Utilisation of Potential

A lot of misconception float around regarding the lag in the utilisation of irrigation potential created and the so called "poor capacity utilisation" of big projects. Full capacity utilisation is a rare phenomenon, even in any industrial or power generation unit which are compact commercial enterprises. The anology of "capacity utilisation" in irrigation projects are kharif irrigation oriented and are basically meant for supporting kharif crops in the event of failure of or delayed rains. Their utilisation in the years of normal rainfall would be low.

Further, the often repeated lag in the utilisation of the created potential is a consequence of the present practice of reporting the statistics. The projected potential is the area that can be irrigated for 75 per cent dependable flow and projected cropping pattern. The actual flows and hence actual potential of the project is bound to vary from year to year and consequently the actual area irrigated annually will vary. Besides, the actual cropping pattern again varies, depending on farmer's choice and prevailing market forces. Hence even when hundred per cent water is utilised, the statistics may show a difference between the potential and utilisation.

Even with all these variables factors and statistical pitfalls which are not understood by those who raise an alarm, it should be noted that the utilisation percentage of major and medium irrigation projects in 1984-85 has been about 84 per cent. Even industrial and power generation units working at this efficiency are rare.

Time and Cost Overrun

Time and cost overrun is not a characteristic of irrigation projects alone. Projects of almost every sector are affected by this due to overall resources constraint and shifting properties. Further, irrigation projects are different from compact and location specific industrial units. They are spread out in large areas where unexpected technical, social and administrative problems crop up very often which lead to revision of designs, shifting of sites litigation etc. Land acquisition is a major cause of delay in almost every project. Besides the normal escalation of costs which come in the wake of delay in execution, the rise in costs of many projects has been due to a number of factors like expansion in scope during implementation, the decision to extended the distribution system up to 5-8 hectare blocks, adoption of higher standards of rehabilitation and environmental protection etc.

Ills due to not taking care of CAD, Drainage etc.

Integrated planning has been made compulsory in the formulation of projects right from 1982 onwards. Soil conservation, catchment area treated, compensatory afforestation and other environmental protection measures, drainage, command area development etc. are now made integral part of project reports. The conjunctive use of surface and ground water in the command, provision of drinking water supply to the population of the neighbourhood are also insisted upon now, in order to made the major projects to work as overall development plans of the area.

Commercial Viability and Losses

It is true that the renvenue from major and medium irrigation projects in most of the states do not meet the operation and maintenance expenditure. This is mainly on account of the fact that the colonial concept of looking at irrigation projects as commercial enterprises has undergone change in the independent India. In the British India, only such projects were sanctioned which with their renenue collections, could meet the running charges and interest on loans and also make profit to the exchequer. The economic viability of the irrigation projects is now being assessed on the basis of the benefits that would be accured at the social and national level rather than from the revenue from water rates. The existing water rates are extremely low on account of socio-political strategy to subsidise agricultural sector. However, the contribution of irrigation projects to National economy is too well known.

ECONOMICS OF MAJOR PROJECTS

Before implementation of major and medium irrigation projects, techno-economic appraisal is carried out. Till a few years back, the practice was to calculate the benefit cost ratio purely in terms of increase in agricultural production in the past-irrigated conditions. Projects having a benefit cost ratio of 1.5 and above are sanctioned excepting in drought prone areas where the projects with a benefit cost ratio of more than 1 are sanctioned. However, from the last two years, apart from the traditional benefit cost ratio of more than 1 are sanctioned. However, from the last two years, apart from the traditional benefit cost ratio analysis, an economic analysis of internal rate of return on the discounted cash flow basis is also carried out. A sub-committee constituted by the Planning Commission under the Chairmanship of Nitin Desai in their report in 1983, recommended that the minimum cut-off rate of internal rate of return should be 9 per cent in general areas and 7 per cent in drought prone, chronically flood prone and hilly areas. In the last two years the schemes processed by the CWC and recommended for approval of the Planning Commission indicate an internal rate of return varying from 12 to 16 per cent. This itself indicates the direct economic benefit of the major and medium irrigation projects. However, these do not take into account some related indirect benefits. The National Council of Applied Economic Research had estimated that these will be atleast additional 30 per cent of the direct benefits.

Quite a few irrigation projects are at present receiving World Bank assistance. The World Bank carried out detailed economic analysis before taking up a project for assistance. Their analysis also indicates an economic internal rate of return of 12-15 per cent. As already pointed out, there are many other economic benefits resulting from these projects which are not taken into account in the economic analysis.

CONCLUSIONS

While it is difficult to evaluate in any precise terms the total quantifiable economic benefits from major dams, it is obvious from the facts presented that the major dams in India have contributed substantially to the growth of the economy. Thanks to this, the years of famine resulting in loss of precious human lives, constant import of food grains, etc. have disappeared. There have been rapid strides in the development of agricultural production which have brought tremendous prosperity to the rural population in irrigated agricultural areas which is also evident from the purchasing power and the standards of living of farmers in the irrigated areas. Besides, these dams have provided a measure of stability to agricultural production as witnessed in one of the worst drought faced by the the country and consequently the effect on the National economy was only marginal. The growth in the industrial sector is again mainly on account of the increased availability of power. It is also proved from the experience of Punjab and Haryana that major dams have been the prime movers for triggering off a chain of economic activities leading to overall prosperity. The same will also hold good for Gujarat and Madhya Pradesh after the completion of the projects taken up on the Narmada recently. Besides, it has to be clearly realised that there is no other way of meeting the increasing demands of agricultural production as well as power generation for the increasing population with the construction of dams. It is high time that the nation realises these facts and the campaign of dis-information which is currently being carried out is stopped in the national interest.

7

Major and Minor Irrigation Works

Cost Aspects of the Controversy

B D Dhawan

It is inappropriate to derive investment norms for irrigation, plan-wise and category-wise, from plan statistics on outlays and irrigation potential. Such norms for minor irrigation works are inherently underestimated. Their comparison with the corresponding norms for major and medium irrigation works give a misleading picture that minor irrigation is a less capital-using option for the Indian economy. Not only the capital saving argument but also the recurring costs argument appears to be tilted against minor irrigation. Thus the case for according priority to minor irrigation development in Indian water planning on grounds of cost economy is not well founded in facts.

INTRODUCTION

By nature water is a very complex resource. As and when its complexities are ignored, or simplified, by the critics of water planning, controversies can easily crop up. In view of the heightened public interest in this vital resource some of these controversies can spill over to public fora like the daily press, weekly magazines and parliament. Given the ample scope for misunderstanding[1] of crucial aspects of this resource, non-issues can become issues of great concern and attention.

---

1. Even for a country like the US, Charles Howe is constrained to remark in his book *Natural Resource Economics* (Wiley, 1979) that water among the natural resources is most misunderstood.

By and large, the controversial issue of promoting minor or major irrigation works -- a variant of this issue is groundwater against surface water irrigation -- is a non-issue. This is for two solid reason. First, our water endowment is not that bountiful as is often portrayed by many public-spirited persons. Even on full development of all our ultimate irrigation potential, we are not likely to extend the benefit of irrigation to more than 60 per cent of our crop land – this percentage is likely to be around 35 for extremely water-short regions like the Deccan plateau, Gujarat and Rajasthan states of western India. Thus we cannot afford to neglect any of our water resources, half of which are in the form of major and medium works category.

The second weighty consideration emanates from a hydrological standpoint. Hydrologically speaking, both groundwater and surface water are inter-linked resources.[2] It is mainly due to such inter-linkage that water planners plead for development of a conjunctive irrigation system, thereby precluding exclusive emphasis on either surface or groundwater irrigation.

In view of the foregoing perception it does not appear to be a matter of great relevance to investigate which of the two options, namely major and minor irrigation, would prove a better proposition in terms of cost-benefit framework. Given the difficulty of resolving the question of comparative costs and comparative returns, one must be wary of all simplistic propositions in this regard. The pitfalls in comparing costs alone of the two categories of irrigation, let alone their comparative augmentation in farm production, are dealt with in what follows. It is hoped that this will go a long way in exploding the myth about minor irrigation, unlike major irrigation, being deemed as a low-cost proposition for the Indian economy. In a way, the author is carrying forward the story of his short paper,[3] in which he conjectured as follows :[4]

2. Some hydrologists consider the interlinkage so close that they refuse to consider groundwater and surface water as separate entities.
3. B. D. Dhawan, 'Water Resource Development: Groundwater versus Surface Irrigation', *Journal of Indian Water Resources Society,* Volume 8, No. 1, January 1988, pp. 49-52.
4. Also, it was hypothesised that the cost advantage of surface irrigation works is wholly offset by the disadvantage on the benefits side. That is, neither groundwater nor surface water irrigation is a cost-effective option, and thus need not be specifically promoted on grounds of costs and returns.

> If costing of the two sources of irrigation is appropriately carried out, surface irrigation works may score decisively over well irrigation, possibly by a margin of 3 : 2.

The outline of the paper is as follows. To begin with, we briefly describe the criteria for categorising irrigation works in Indian planning. Next we look at the cost structure of irrigation. This is to emphasise the need for probing the cost issue both from capital and recurring cost angle, as the comparisons done by some writers solely or the basis of capital costs unfairly show major irrigation works in a poor light. This kind of partial comparison of the two categories of works is still further tilted against major irrigation as the available estimates of unit capital cost are found to be deficient on three counts. The drawbacks of the available estimates, derived mostly from five-year plan reports, are duly highlighted. This is followed by drawing attention to the weakness of the much-talked about capital saving argument in favour of minor irrigation development. Towards the end, we bring out the substantial advantage in respect of recurring costs in promoting major over minor irrigation in India.

Since neither our estimates of unit capital cost nor our estimates of operational costs of major and minor irrigation works are very firmly established, we are desisting from indicating the extent to which the overall cost advantage lies in favour of major instead of minor irrigation works. Whosoever is eager to do so must use not only firm estimates of capital and operational costs but also carry out the exercise in terms of social costs. While doing so, one must duly reckon with two aspects of investment costs of major irrigation works, namely, (i) the true cost of land (including forest cover) submerged by reservoirs an d (2) the rent-seeking elements (i.e., kickbacks and fictitious expenditure) in costs of major irrigation projects. While the true cost of land submergence is likely to be far more than the nominal figures shown in project records, the exclusion of rent-seeking costs will substantially bring down the true resources cost of the major irrigation projects.

CATEGORIES OF IRRIGATION WORKS

Irrigation works can be distinguished in more than one way. For our purpose here it would suffice to note that categorisation of irrigation works in Indian planning is mainly on the basis of irrigation capacity of a work, mentioned as irrigation potential in our plan documents. (Prior to the Fifth Five-Year Plan, investment outlay needed for executing an irrigation work was the criterion of categorisation). When the crop area to be irrigated from a given scheme is 2,000 hectares or less, the scheme is treated as minor irrigation work. It is when the irrigated acreage exceeds the limit of 10,000 ha that the irrigation works is characterised as major work. For a scheme designed to irrigate between 2,000 and 10,000 ha there is the third category of medium irrigation works. Schemes under the third category being few, these are usually clubbed with the category of major irrigation works. And all groundwater-based schemes are invariably included in the minor irrigation category.

Leaving aside for the moment the limitations of plan statistics with regard to major and minor irrigation potential, the aggregate irrigation capacity of all minor irrigation works is the order of 40 million ha as compared to about 30 million ha under major (and medium) irrigation works. Thus the minor irrigation segment of the Indian irrigation sector is bigger than the major irrigation segment by a factor of one-third.[5]

Within minor irrigation segment, dugwells and tubewells dominate, accounting for almost three-fourth of the capacity of this segment. In fact, most of the additions to minor irrigation capacity since 1951 are from groundwater sources.

COST STRUCTURE OF IRRIGATION

For any productive activity, two types of costs are entailed. One is the capital cost, and the other is the operational cost. Capital costs arise in the construction phase, whereas operational costs are incurred after the construction phase is over. It is now well recognised that modern irrigation works are highly capital-using. In view of this, project cost comparisons are often made in terms of unit capital costs of various irrigation projects. That is to say, irrigation projects are compared on the basis of their construction costs (including pre-investment inves-

5. B. B. Vohra, a proponent of the minor irrigation cause, is unduly dramatising by writing under the banner 'Minor becomes Major'. In point of fact, such a differential between the two segments has prevailed since long, e.g., the same differential is noticeable in 1950-51, the pre-plan date.

tigational outlays) per unit area to be irrigated by them on full development. This way of looking as comparative project cost appears quite all right for major irrigation works whose annual operational costs (including repair and maintenance costs) per unit area to be irrigated are extremely small. According to the Eighth Finance Commission of India, the operational costs for major irrigation works for 1981-82, when divided by area benefited by these works, would amount to about Rs 92 per irrigated hectare.

Evidently, it is inappropriate to compare two irrigation projects on the basis of unit capital cost only if their cost structures are substantially at variance with each other. This is precisely the position when it comes to comparing major with minor irrigation works. It is a well known fact that operational costs vis-a-vis capital costs are of far greater significance for minor than for major irrigation works. Given this structural difference in costs, cost comparison based on capital cost alone can put minor irrigation in a questionable advantage over major irrigation. And matters can become truly worse if the estimates of unit capital cost of the two categories of irrigation are highly deficient, the deficiency being such as to tilt the balance in favour of minor irrigation as shown below.

DRAWBACKS IN AVAILABLE ESTIMATES

Nowhere in the five-year plan reports one comes across concrete figures of unit capital requirements for irrigation works of either category. Therefore, one must be wary of any such estimates that may be quoted by minor irrigation lobbyists in support of their contention. The cited figures, category-wise or plan-wise, are usually the handiwork of analysts who try to grind them out from the following two sets of statistics listed in the plan documents:

(1) Actual or anticipated outlays for major and minor irrigation works for each plan period.
(2) Planners' own estimates of irrigation potential created, or anticipated to be created, in each plan period for each category of irrigation.

In what follows we bring out how faulty estimates of unit capital cost can be generated when plan outlays are divided by corresponding estimates of irrigation potential. To begin with, the plan statistics on outlays for minor irrigation do not give a full picture of capital resources required for developing minor irrigation capacity in the country. The reported outlays are for the public sector only, thereby leaving out investments on private account. In other words, the plan outlays for minor irrigation works are not inclusive of farmers' own investments in tubewells, dugwells, pumpsets etc. Since the plan statistics on minor irrigation potential are for the entire minor irrigation segment (i.e., both on public and private account), the collation of the area and outlays data given in the plan documents gives rise to seriously underestimated figures of unit capital requirement of minor irrigation unless the user of these plan statistics makes suitable adjustments for the omitted investment on private account. Needless to say, a similar collation for major and medium irrigation works yields an estimate of unit capital cost which is free from the element of underestimation as the major and medium irrigation schemes are envisaged entirely for the public sector.

Secondly, there is some lack of compatibility between outlays on irrigation in a given plan period and the resultant addition to irrigation capacity in the same plan period. It is well known that the gestation period for a major irrigation project is about 10 to 12 years under efficient conditions of implementation, and 20 to 25 years under inefficient conditions of implementation. As irrigation capacity of an ongoing project is not created in direct proportion to investment expenditures in a given plan a lack of correspondence between plan outlays and created irrigation potential can exist for a given plan period. This can distort the estimate of unit capital cost, the distortion being more serious in the case of major than minor irrigation works like wells which can be usually constructed with a single year.

The third limitation of the plan outlays on irrigation arises from the fact that some of these are not meant to result in any addition to irrigation capacity. For instance, outlays on (1) research and investigation organisation like the Central Water Commission and Central Groundwater Board; (2) remodelling of headworks, lining of channels and strengthening of drainage in old irrigation and (3) replacement of old works[6] that have outlived their useful lives, are not envisaged

6. A sum of Rs. 8,000 million is to be spent as replacement investment on the Ganges Canal System (west UP) whose major civil works (between Roorkee and Haridwar) are now more than 125 years old. The World Bank is financing this scheme. Likewise, Rs. 2,000 million is the replacement investment needed for the Sone Canal of south Bihar.

to enhance irrigation capacity as such. Instead, these are meant more to improve the productivity of the irrigation system as such. Evidently, such outlays unduly raise the derived figures of unit capital costs, the rise being obviously higher for major works.

The foregoing exposition of the principal drawbacks of the plan data on irrigation sector clearly brings out that the estimates of unit capital cost for irrigation works, when derived from such data, can be quite erroneous. Such derived estimates, as we shall demonstrate presently, could be grossly understated in respect of minor irrigation works--especially for wells and tubewells– and considerably overstated for major and medium irrigation works. Evidently, inferences based on a comparative analysis of such unit capital cost estimates for various irrigation works cannot be accepted without reservation. In his recent book, J. S. Kanwar, basing himself on such faulty estimates, erroneously observes that the 'cost of groundwater development are hardly one-fourth of surface water development in the Seventh Plan'.[7] Likewise, in another recent study by the secretariat of the Afro-Asian Rural Reconstruction Organisation, the following inference is drawn for the entire period of Indian planning:

> The investment cost of unit irrigated area for major and medium irrigation projects was 118 per cent more than for minor irrigation.[8]

For the benefit of the curious reader, we give in the table the estimates derived by P. Sachindanandan and B. P. C. Sinha for major and minor irrigation works for various plan periods.[9] Mainly owing to the persistence of price inflation throughout the period of planning, unit investment cost for all categories of irrigation works have shown a strong upward rise, notably since the Third Five-Year Plan period. During the fifties this rise was rather of a modest magnitude in the

7. J. S. Kanwar (ed), *Water Management – The Key to Developing Agriculture*, New Delhi, 1988, p. 7.
8. *Source:* 'Role of Irrigation in Development of Agriculture in Developing Countries', Rural Reconstruction, Volume 20, No. 1, January 1987, p 82. See also the subsequent issue of this journal.
9. It may be pointed out here that separate plan outlays for groundwater development are not listed in the plan documents. It is by assuming groundwater development to be synonymous with minor irrigation development that writers like J. S. Kanwar and B. B. Vohra argue about groundwater being more cost-effective than surface irrigation in India.

category of major and medium irrigation works, but very steep in the category of minor irrigation works (see the Table 7.1). As a result of this, the Second Plan figure for unit cost for minor irrigation works came to exceed the corresponding figure for the major and medium irrigation works. Our contention is that this startling result has arisen largely because of serious underestimation in the estimated outlay figure for minor irrigation for the First Plan period.

Institutional finance has played a note-worthy supporting role in the development of minor irrigation. Over time, its share in total investment funds used for minor irrigation works has risen steadily, though firm estimates in this regard are not available. This is inferred from two empirical facts. First, investment in Indian agriculture are being increasingly financed by the commercial and the co-operative banks. According to M. V. Gadgil, institutional finance, which accounted for 29 per cent of private capital formation in Indian agriculture in 1973-74, constituted about 43 per cent of capital formation during 1982-83.[10] Second, much of this institutional finance has been used for minor irrigation purpose. According to S. S. Sangwan, the per cent share of such finance was 42 in 1969, 80 in 1975, 63 in 1981 and 49 in 1985.[11] In view of these facts, one may say that the degree of underestimation in investment outlays for minor irrigation ought to diminish as we move from First to the Seventh Plan.

QUESTIONABLE CAPITAL SAVING ARGUMENT

In a capital-short economy like ours it is a very appealing thing to commend to planners options which appear least capital using. In this context, we must reckon with two things, namely, (1) the supporting investments required for each option and (2) the replacement time after which fresh investment is required for each option. Let us, therefore, look at these aspects of major and minor irrigation options, so as to ascertain which option may strain our capital resources less in the long run. In my assessment, the capital saving argument, contrary to the

---

10. M. V. Gadgil, 'Agricultural Credit in India: A Review of Performance and Policies', *Indian Journal of Agricultural Economics*, Vol 42, No. 3, Conference Number, Part I, July-September 1986, p 293.
11. S. S. Sangwan, 'Agricultural Investment and Regional Imbalances. A Study of Refinance Disbursed by NABARD', *Indian Journal of Agricultural Economics*, Vol 42, No. 3, Conference Number, Part II, July-September, 1986, p 567.

general impression, is more tilted in favour of major than minor irrigation.

Generally speaking, the useful life of major irrigation works is reckoned at 100 years, after which the civil structures of the canal network need to be reconstructed. As far as the dam is concerned, its life is governed mainly by the number of years it would take to get the 'dead' storage of the reservoir to fill up with silt and debris brought by water flowing into the reservoir. It is widely agreed that the rate at which reservoir siltation is occurring in India is much higher than the rate assumed by dam designers. Even then, reservoirs are expected to serve useful purpose for 100 years of thereabouts. So, the need for replacement investment in the case of major irrigation works is anticipated to materialise largely after 100 years. This is far in excess of the corresponding period for minor irrigation works, especially tubewells which are expected to last for 20 years or so under good maintenance conditions. Likewise, pumpsets fitted on open wells need replacement investment in 15-20 years time. It is only the investment embedded in dugwell proper that requires replacement after a much longer time, say, about 50 years in the case of masonry wells located in alluvial soils and 100 years or more for dugwells in hard rock regions.

Major and medium works use the natural force of gravity in their operation, while groundwater and minor surface water lift schemes need energy for countering the pull of gravity. Thus supporting investments in the energy sector are minimal for major and medium works but quite substantial for minor irrigation. In particular, electric pumpsets/ tubewells need requisite investment in power generation equipment and rural electrification system. An average pumpset in Indian required about 4 KW of power during the mid-seventies when each kW of generating capacity cost nearly Rs 3,000 in capital outlay alone.

It is not known how much is the capital cost borne by the nation while meeting the energy needs of an agricultural pumpset. Nor do we have firm estimates of such supporting investment needed for each unit area of irrigation potential established through minor irrigation works. But some idea about the relative dimension of such costs vis-vis the plan outlays on minor irrigation can be formed by adopting the following procedure, which we apply to the Fifth-Five-Year Plan data. To begin with, we estimate that portion of the plan outlay on power generation as can be attributed to 0.870 million electric pumpsets added during the same plan period. To this we add the specific plan outlay earmarked for rural electrification. Finally, we estimate the investment in minor irrigation on farmers' own account. Adding these three costs on capital account to the direct capital outlay reported for minor irrigation, we obtain total investment needs of the minor irrigation segment, which on division by the corresponding addition to minor irrigation potential yields the estimate of total investment cost per unit area for minor irrigation works.

Assuming (1) each electric pumpset requires 4 kW capacity at the power generating point and (2) each kW capacity was established at Rs 3,5000 during the 5th Plan, the plan outlay for power generation for 0.870 million electric pumpsets would amount to about Rs 12,180 million. To this may be added Rs 8,420 million on account of expenditure on rural electrification programme during the Fifth Plan, making a total of Rs 20,600 million on account of supporting investment in power for electric pumpsets.[12]

As against the supporting, or indirect, investments of the order of Rs 20,600 million on account of electric pumpsets, the Fifth Plan figure of direct investment in minor irrigation is considerably lower, namely, Rs 14,110 million. As pointed out earlier, this figure is exclusive of investment on farmers' own account. The reported break up of this figure is as follows: (i) contribution of the state government Rs 6,310 million and (ii) financing from institutional funds Rs 7,800 million.

If we make an assumption that farmers' own funds amounted to just *one-fifth* [13] of the total direct investment in minor irrigation, the farmers' share would be around Rs 3,528 million. Thus aggregate direct oulays on minor irrigation for Fifth Plan add up to nearly Rs 17,638 million, still less than the corresponding indirect outlay of RS 20,600 million on

12. Admittedly, there is overestimation here because of our tacit assumption that the generating capacity of 4 kW and the rural electrification system are wholly and exclusively for electric pumpsets. But then, we are also not reckoning with supporting investments for diesel pumpsets.

13. The Working Group for Sixth Plan assumed that 35 per cent of the outlay on private wells would be financed by farmers themselves (p 17). But if we reckon with total investments of minor irrigation, the share of private investment is Rs. 1,000 crore out of Rs. 4,850 crore, i. e., about one-fifth.

account of electric pumpsets. The sum of these direct and indirect outlays comes to about Rs 38 billion, which on division by 3.8 million ha of minor irrigation potential added during the Fifth Plan yields a estimates of nearly Rs 10,000 per ha. This re-estimated unit capital cost is well above the corresponding estimate of nearly Rs. 6,000 per ha for major and medium irrigation works (*vide* the table).

Table 7.1 : Unit Investment Figures for Irrigation by Type of Irrigation and by Phase of Indian Planning

S No	*Planning Phase*	*Major and Medium Irrigation Works*	*Minor Irrigation Works (A)*	*Minor Irrigation Works (B)*
1	First Five-Year Plan (1951-56)	1,520	550	655
2	Second Plan (1956-61)	1,810	2,300	2,210
3	Third Plan (1961-66)	2,640	2,010	2,820
4	Annual Plans (1966-69)	5,900 (2,890)	2,810	1,990
5	Fourth Plan (1969-74)	4,760	2,610	2,610
6	Fifth Plan (1974-78)	6,110	2,710 (3,710)	3,670
7	Annual Plans (1978-80)	10,820	4,120 (3,620)	3,660
8	Sixth Plan (1980-85)	19,310	4,520	4,390
9	Seventh Plan (1985-90)	26,870	7,830 (7,330)	NA

Notes : (1) For major and medium works as well as minor works (estimate A), the data are from Sachidanandan's paper. Some discrepancies were noticed. So the correct figures, as derived from his data on expenditure and irrigation potential given in Table 1 (p 80), are shown in brackets.

(2) The estimate B for minor works is derived by the author from Sinha's relevant data in Table 6 (p 140). For Sixth Plan, the estimate is based on provisional data.

Source : Papers by P. Sachindanandan and B. P. C. Sinha in J S Kanwar's book.

DIFFERENTIAL IN OPERATIONAL COSTS

The weakness of the assertion that, hectare for hectare, minor irrigation requires a fraction of the capital resources needed for developing major irrigation works has been demonstrated in the preceding analysis. In all probability, establishment of major irrigation projects may prove a much less capital using proposition for the economy as a whole. In what follows an attempt has been made to quantify the advantage in recurring costs in promotion major over minor irrigation schemes. What is being attempted here is an all-India picture of recurring as well as one-shot costs of the two broad categories of irrigation works. It is very likely that the differential in costs varies considerably across regions of this country of continental diversity.

The estimational details, alongwith the sources of data and their limitations, are not being enumerated here in the text but are being relegated to an appendix note. Here, two observations would suffice for a discerning reader. First, the operational costs are strictly recurring costs, thereby excluding both the depreciation charges as well as the interest charge on investment cost. Second, an attempt has been made to work out the recurring cost in resource use terms. While we have tried to keep out the elements of taxes and subsidies, no attempt has been made to shadow price the inputs. Thus, the cost of electricity used in wells equipped with electric pumpsets (now the most prevalent mode of groundwater irrigation in the country) is reckoned at the resource cost of supplying it to rural areas. That is, we value electricity neither at its opportunity cost elsewhere in the economy, nor at the highly-subsidised rates actually paid by farmers.[14]

Obtaining estimates of recurring costs for major and medium irrigation works is a comparatively simple task as the comptroller and auditor general of India compiles financial accounts of public irrigation works in one consolidated form. On the other hand, it entails considerable effort in estimational work for minor irrigation works, which are largely owned by individual farmers. We have attempted the exercise for the beginning of the eighties only. Needless to say, there is ample room for improving the estimational work.

Our estimate for major and medium irrigation works comes out to nearly Rs 60 per irrigated hectare. This is way below the corresponding estimate of recurring expenditure for groundwater

14. For a 5 HP pumpset, the price realised by the public utilities is estimated to vary widely from state to state. It is about Rs. 0.15 per kWh in Kerala, MP and Maharashtra; between Rs. 0.10 and Rs. 0.15 per kWh in Andhra Pradesh, Jammu and Kashmir, Karnataka, Punjab and Tamil Nadu; between Rs. 0.15 and Rs. 0.25 per Kwh in HP, Orissa, Rajasthan, Chandigarh, Delhi, and Pondicherry; between Rs. 0.25 and Rs. 0.35 per kWh in Assam, Haryana, Meghalaya, West Bengal and Tripura; and between Rs. 0.35 and Rs. 0.45 in Bihar, Gujarat and UP (Central Electricity Authority, *Average Electric Rates and Duties in India*, 1986, p 15).

irrigation, which ranges between Rs. 637 and Rs. 849 per irrigated hectare. Thus there emerges a differential of almost 1 : 10 between major and minor works, that is, major irrigation is one-tenth less resource using in respect of recurring costs.

CONCLUSIONS

It is inappropriate to derive investment norms for irrigation, plane-wise and category-wise, from plan statistics on outlays and irrigation potential. Such norms for minor irrigation works are inherently under-estimated. Their comparison with the corresponding norms for major and medium irrigation works give a misleaing picture that minor irrigation is a less capital using option for the Indian economy. Not only the capital saving argument but also the recurring costs argument appears to be tilted against minor irrigation. Thus the case for according priority to minor irrigation development in Indian water planning on grounds of cost economy is not well founded in facts.

APPENDIX

ESTIMATES OF RECURRING COSTS OF IRRIGATION

In this note we spell out the details behind our estimates of recurring costs of major and minor irrigation per irrigated hectare for the beginning of the eighties. There is considerable room for firming up these estimates, especially the one for minor irrigation. Despite the infirmities, we do hope these numbers do provide a fair picture of the differential in recurring costs between the two categories of irrigation works.

SOURCES OF DATA AND THEIR LIMITATIONS

The annual reports of the state irrigation departments are the principal source of information with regard to operational expenses of major irrigation works. The audited accounts of these reports are published by the comptroller and auditor general of India in its annual publication entitled *Combined Finance and Revenue Accounts of the Union and State Governments of India*. Accounts of major and medium irrigation works figure in two sets of tables, namely (a) Account No 101 for multipurpose river valley projects like Bhakra and Nagarjunasagar, and

(b) Account No 102 for irrigation, navigation, drainage and flood control projects. While the data in respect of irrigation is separately listed in Account No 102 (and its sub-accounts), it is not so with the other account. Expenditure on two items, namely, maintenance and administration, is being deemed by us to be the operational cost of major irrigation works.

For getting at the operational costs of minor irrigation, one has to carry out an estimational exercise. The task is not difficult because we have adequate data base for the purpose. First, we have the published records on electricity consumed by the Indian agricultural sector. Since the resource cost of supplying electricity can be computed from the records of the state electricity boards, we can make a reasonable estimate of operational cost of well irrigation due to the use of electricity. Second, we have published records on the number of waterlifting devices by category of motive power, namely, electric pumpsets and diesel pumpsets. By adopting a suitable norms for petroleum oil consumption per diesel pump per year, one can figure out consumption of diesel oil for well irrigation purpose, and this consumption can be valued at appropriate price of diesel oil. Third, repair and maintenance cost of pumpsets can be figured out by first adopting an investment norm per pumpset, and next applying the norm that repair and maintenance expenditure amounts to some 5 to 10 per cent of the cost of pumpset. Fourth, operational costs for wells equipped with traditional waterlifts may be assessed with the help of estimates based on sample survey reports prepared by scholars/ agencies who have studied the comparative economics of traditional and modern wells.

A word about the data source on irrigated area is necessary. We are using the data on utilised irrigation potential given in five-year plan reports, and not from the land use statistics compiled by the state revenue departments and finally published by the central ministry of agriculture. The problem with the latter source, which is usually used by scholars of Indian agriculture, is that category-wise statistics of irrigated area are in 'net' terms, and not gross terms. Without making assumptions about intensities of irrigation for major and medium irrigation works on the one hand and minor works on the other, it is not possible to convert net irrigated area into its corresponding gross irrigated area. Fortunately the plan statistics of irrigation potential are on gross area basis.

OPERATIONAL COST OF MAJOR IRRIGATION

The operational expenditure for the multipurpose river projects amounted to about Rs. 279 million during 1980-81. In respect of major and medium irrigation projects the same totalled around Rs. 1,083 million. Thus overall operational costs of canal irrigation may be reckoned at Rs. 1,362 million, of which Rs. 530 million was accounted by establishment cost alone. The maintenance and repair expenditure stood at about Rs. 844 million, just about 15 per cent below the revenue income (Rs. 986 million) from the canal system in 1980-81. Adopting a figure of 23 million hectare for area actually irrigated by major and medium works in 1980-81, one obtains an approximate estimate of Rs. 60/ha as operational cost for canal irrigation.

OPERATIONAL COST OF WELL IRRIGATION

There were about 4 million electric pumpsets at the beginning of the financial year 1980-81. Indian agriculture reportedly absorbed in this year about 14,489 million kWh. These statistics imply an annual consumption of about 3620 kWh per electric pumpset, which is about one-fifth higher than the norm of 3000 kWh per electric pumpset used by the Working Group of Energy Policy in its report of 1979. Assuming that some of the electricity was used for purposes other than pumping of groundwater, we are envisaging that about 13,000 million kWh was used for groundwater pumping. Though farmers usually paid 15 to 20 paise per kWh, the resource cost of power generation and rural electrification in 1980-81 may be placed in the range 60 to 80 paise per Kwh.[15] Thus, cost on account of electricity comes in the range of Rs. 7,800-10,400 million.

Diesel oil used in agriculture, as per the national income accounts of 1980-81, amounted to about Rs. 24,788 million at current prices. This was used by over 2.6 million diesel pumpsets, nearly 0.3 million tractors, an unknown number of harvest combines and threshers. The Working Group on Energy Policy (1979) adopted a norm of 0.8 tons per diesel pumpset. (We believe that this is on the high side, as it is derived on the principle of energy equivalence for 3000 kWh for an

15. For Uttar Pradesh, Rajadhyaksha Committee on Power (1980) found that resource cost of supplying electricity to rural areas during 1978-79 amounted to about 67 paise/kWh as against the average cost of 44 paise for all LT consumers.

electric pumpset.) If we follow this norm and assume a price of Rs. 2,200 per ton for diesel oil, the energy cost for diesel irrigation pumps may be placed at Rs. 4,576 million for 1980-81. But if we adopt a lower norm, say 0.6 tons per diesel pumpset, the cost for 2.6 million diesel pumpsets would be Rs. 3,432 million. Thus we get a range estimate of Rs. 3,432-4,576 million for diesel consumption by diesel pumpsets for 1980-81.

Thus the expenditure on account of electricity and diesel oil fell in the range Rs. 11,232-14,976 million, which, when divided by a groundwater irrigation potential of 23 million ha for 1980-81, yields a range estimate of Rs. 488-651 per ha. To this may be added a notional repair and preventive maintenance cost of Rs. 49-98 per ha by making two assumptions, namely, (1) the investment cost per pumpset in 1980-81 was Rs. 3,500 and (2) the repair and maintenance cost ranged between 5 and 10 per cent of the investment in pumpset. Therefore, the operational cost on account of pumpsets along ranged between Rs. 537 and Rs. 749 per irrigated hectare. To this have to be added two more costs, namely, (i) cost of operating traditional wells and (2) repair and maintenance of dugwells and tubewell borings.

There were about 7.8 million dugwells and nearly 2.2 million tubewells in existence at the beginning of 1980-81. If 6.5 million pumpsets were used in 2.2 million tubewells and 4.3 million dugwells, 3.5 dugwells are presumed to be operated by traditional waterlifts. Since many of the dugwells fell into disuse as their owners acquired tubewells, we may say about 2.3 million dugwells were used with traditional waterlifts. We are adopting a figure of Rs. 1,000 per traditional waterlift.[16] With this, the average operational cost per hectare for well irrigation stood somewhere between Rs. 637 and Rs. 849 during 1980-81.

16. That our figure of Rs. 1,000 is not unduly high is sustained by the following comparative picture of operational costs for different waterlifts for west UP for 1984-85.

Rope and bucket lift ('mhot')	= Rs 2331
Counterpoise lift	= Rs 963
Persian wheel	= Rs 2494
Diesel pumpset TW	= Rs 3335 - 5322
Electric TW	= Rs 1561 - 2202

(*Potential of Small Scale Lift Technologies with Special Reference to Access of Small and Marginal Farmers to Ground Water.* Water Resources Development Training Centre, University of Roorkee, Chapter VI (draft report).

COMMENTS

Ashok K. Mitra

The tone, tenor and contents of Dhawan's paper entitled, 'Major and Minor Irrigation Works : Cost Aspects of the Controversy' (EPW, September 30, 1989) are indeed intriguing. With this paper he appears to have switched roles from an unbiased social scientist to a crusader of big dams. From time to time Dhawan has been writing on the hotly debated issue of promoting minor or major irrigation works in India. His stand on the issue, till the piece under discussion saw the light of the day, appeared to be unbiased and based on sound analytical framework and logic, and being a specialist on the subject matter of Indian irrigation development, he has been advocating that development of major surface irrigation system alongwith groundwater exploitation is a must for agricultural growth and development in India. But, unfortunately his latest paper on this issue gives one the impression that he is betraying the symptoms of a lobbist for the cause of big dams.

As a matter of fact he begins even this paper on a sober note and with a plea that by and large the controversial issue of promoting minor or major irrigation works is a non-issue and that considering the endowment of our water resources, which certainly is not bountiful, we cannot afford to neglect any of our water resources. But, thereafter he suddenly sets out to demolish the arguments of cost effectiveness of minor irrigation works with a penchant with data which are not only questionable but are very scanty and on arguments based on very weak analytical framework and premises. His entire argument in favour of cost effectiveness of major and medium irrigation works is based on the comparison of unit total cost of irrigation from major and minor (only groundwater) sources. He emphasises with a great deal of conviction that to examine the cost effectiveness of various sources it is necessary to look into the operational cost also in addition to the capital cost and that cost comparison based on capital cost alone can put minor irrigation in a questionable advantage over major irrigation. Based on his analysis (the weakness of which we shall presently show) he concludes that not only the capital saving argument but also the recurring cost argument appears to be tilted against minor irrigation. Thus, the case for according priority to minor irrigation development in Indian water planning on grounds of cost economy is not well founded in facts.

Before we take up the data used and methodology followed by Dhawan in arriving at such dubious conclusion, it is necessary to make a few observations on the comparison of this type. On Dhawan's own admission water is a very complex resource and does not lend itself to any simplistic analysis, and any attempt to do so may give rise to questionable conclusions and hence controversies. Yet, Dhawan, in this paper has exactly done so disregarding his own caution. Precisely because the cost structures are substantially at variance with each other in respect of major and minor irrigation works simplistic type of comparison which Dhawan attempts would invite more controversies despite Dhawan's claim of having taken into consideration the unit operational cost in addition to unit capital cost. Notwithstanding the weaknesses of such comparisions since Dhawan has attempted one such, in what follows, we discuss the data inadequacy and weakness of the methodology of even such a comparision as has been attempted by him.

Although Dhawan's exercise is not carried out in terms of social costs he claims that even after reckoning with two aspects of investment costs of major irrigation works, namely, (i) the true cost of land (including forest cover) submerged by reservoirs, (ii) rent seeking elements in costs of major irrigation projects, on balance the true resource costs of major irrigation projects will be substantially reduced on account of the latter component. One fails to understand how Dhawan arrives at such a conclusion. On the contrary, the rent seeking elements (i.e., kickbacks and fictitious expenditure) which are to be excluded from total cost would be peanuts before the real resource cost in terms of land submerged and loss of forest by major irrigation works and to this if we add real resource cost of rehabilitation of the displaced households the real resource cost would appear to be phenomenal.

Let us now turn to the main arguments of Dhawan's paper, namely (i) drawbacks in available estimates of capital cost, (ii) questionable capital saving argument and (iii) differential in operational costs. To begin with one would readily agree with Dhawan that any estimate of unit capital cost of either category is suspect because of weakness in the basic data on actual or anticipated outlays for major and minor irrigation works for each plan period and on planners' own estimates of irrigation potential created or anticipated to be created in each plan period for each category of irrigation. But then Dhawan's contention that these drawbacks of plan data on irrigation lead to gross underestimation of unit capital cost in respect of minor irrigation works, to say the least, is untenable. His first point that the reported outlays in plan statistics in the case of minor irrigation are for the public sector only is well taken and hence before any unit cost is estimated, farmers' own private investment in tubewells, dugwells, pumpsets, etc., would have to be added and to that extent the unit capital cost for minor irrigation works would increase. His second point that the lack of compatibility between outlays on major irrigation in a given plan period and resultant addition to irrigation capacity in the same period because of long gestation period can distort the unit capital cost against the cause of major irrigation works can not be substantiated and is questionable. Indeed if the gestation period of major irrigation work is anywhere about 10 to 25 years interest on capital investment yearwise ought to be capitalised until the project is fully developed and commissioned for use and this would certainly increase the unit capital cost of major irrigation works. His third point that outlays on remodelling of headworks, lining of channels, strengthening of drainage and replacement of headworks do not result in any addition to irrigation capacity but their inclusion in the total outlays during the plan period increases the unit capital cost for major irrigation unnecessarily can also be counterargued. It is well known that the major irrigation systems work with much reduced efficiency. It is only because of the correction of the initial faults in design and distribution network with these outlasy that the bulk of the potential created is actually irrigated. Though outlays on these account appear with a lag, in our opinion, these are very much a part of total capital outlay and can not be dispensed with without affecting the potential created or anticipated to be created.

Let us now turn to questionable capital saving argument raised by Dhawan. It is strange that suddenly he changes his focus from plan period capital outlay and corresponding additional irrigation potential created under major and minor irrigation works to ascertaining which option may strain our capital resources less in the long run. So, naturally, he raises the issue of the replacement time after which investment is required for each option. While doing so he argues that reservoirs under major irrigation works are expected to serve useful purpose for 100 years or thereabouts and hence the need for replacement investment in the case of major irrigation works is anticipated to materialise largely after 100 years. This argument is even more strange considering that he himself admits that the rate at which reservoir siltation is occurring in India is much higher than the rate assumed by dam designers and yet he questions the wisdom of including the outlays on remodelling of headworks, lining of channels and strengthening of drainage in old irrigation and replacement of old works in the capital cost when all of these are incurred to maintain the productive capacity of the major irrigation system in the long run.

The most shocking analysis and argument of Dhawan appear in the portion dealing with estimation of supporting or indirect investment needed for each unit area of irrigation potentials established through minor irrigation works. Nothing can explain a figure of Rs. 20,600 million as indirect investment attributed to agricultural pumpsets other than figment of Dhawan's mind enraged at being constantly told about the cost effectiveness of minor irrigation works. Data presented on this account are so vague and scanty and the assumptions involved in the estimates are so far fetched that one would like to dismiss these absurd estimates at the first sight. Indeed Dhawan, a rational analysist and an eminent scholar that he is, himself is aware of the pitfalls in his estimates, but, perhaps in his eagerness to prove his point, decided to relegate the critical observations on his estimates only in the footnote, when these observations ought to have desisted him from attempting such an estimate. One may also raise questions on the authenticity of the data used in such estimates of crucial importance. Further, it is also not clear what constitutes an expenditure on rural electrification; is it on account of transmission and distribution or also includes expenditure on account of power generation ? One would also like to know what are the supporting investment for diesel pumpsets which Dhawan has not so graciously reckoned with, as mentioned in the footnote. This argument can also be extended to the absurd level of reckoning with the supporting investment for creating major irrigation works.

Finally, we come to the question of operational cost of well irrigation. We wonder if the cost of generation of power for agricultural use (even though erroneously estimated) and the cost of transmission and distribution (supply) through investment in rural electrification is already taken into account while estimating the unit capital cost (as has been presumably done by Dhawan), what is it that Dhawan is estimating as operational cost other than repair and maintenance in respect of electrically energised wells ? If he is also adding cost of power used to the above, would it not amount to double counting ? In the appendix of his paper Dhawan presents the details of the estimation of operational cost of well irrigation. We are referring to the electrically energised pumpsets only and according to Dhawan's own estimate the resource cost of power generation and rural electrification in 1980-81 may be placed in the range 60 to 80 paise per kWH. Thus, Dhawan states that cost on account of electricity comes in the range of Rs. 7800 - 10,400 million. If this is the resource cost of power generation and rural electrification then in what account Rs. 20,600 million is also attributed to investment on power generation and supply under the head of capital cost ? Apart from infirmities, there surely appears to be a double counting of serious dimension in Dhawan's estimate.

From the above discussion it would appear that even with the simplistic comparison attempted by Dhawan the case for according priority to major irrigation development in Indian water planning on grounds of cost economy is not established.

It may also be mentioned that capital cost on major irrigation systems on account of many additional works, such as, construction of field channels, land levelling, drainage, etc. which need to be carried out before the farmer can irrigate his field has escaped Dhawan's attention. If these costs are added the final figure of unit cost on major irrigation works would be much higher than what is estimated by Dhawan. There is then the cost involved in the incalculable human suffering caused by uprooting communities from their ancestral habitat and place of work and resettling them elsewhere in a manner which is never fully satisfactory. In view of the above facts the question that arises is should the planners remain preoccupied with large and medium storage route to irrigation development alone which may be fraught with unfavourable consequences and does not always take into consideration conservation of natural resources, or should they also look for other alternatives ? Perhaps it is time that we pondered over these issues very seriously and before it is too late suspended the disproportionate efforts towards big storage dams for the time being.

REPLY

B D Dhawan

I welcome Ashok K. Mitra's observations to the extent these pertain to the contents of my paper. These do merit attention, especially by upcoming scholars who may wish to ascertain the true costs of Indian irrigation, more so the comparative cost differential between minor and major irrigation. In this rather ticklish pursuit, they should not be daunted by reputations of their seniors in the profession; here they need to emulate Mitra in questioning closely the assumptions of a 'rational analyst and eminent scholar'.

Contrary to the general impression, there is ample room for difference of opinion/assessment in the simple matter of costs of an activity. Costing is not free from speculation. While many analysts are readily willing to admit of the presence of speculative elements in the benefits side of a cost-benefit exercise, few do so in respect of the cost side. It is only when one fully grasps the significance of adages, like "costs and benefits are two sides of the same coin", or "benefits are nothing but costs averted", that one realises the speculative nature of cost exercise also. At surface, Mitra's differences with me may appear to be one of those proverbial cases of economists' being at loggerheads with one another (they do occasionally cooperate and collude in "academic empire" building). But at depth the core of these differences are genuine to the extent these are due to 'figments of the mind'.

Take, for instance, his assertion that leakages in irrigation investments would be peanuts in comparison to the real costs of land submergence and deforestation due to big dams. In this assertion he is at best guided by his experiences with Maharashtra irrigation. Experiences elsewhere may be different, and more importantly, may be perceived differently for the same state. To illustrate, while one senior bureaucrat from Uttar Pradesh told me in all earnestness that 15 to 20 per cent may be the extent of leakages in U.P. irrigation works, a family of irrigation contractors in that state placed this percentage in the range of 45-50 in respect of works executed by them for the U.P. Irrigation Department.

The central concern of my paper was not with the estimation of the exact margin of cost differential between major and minor irrigation in India. Such an exercise in a country of continental dimensions be better done region - wise, as has been done by Kannan and Pushpagadan (1989) for Kerala.* I wish that Mitra had apprised us with some of his own estimates about western Maharashtra region on which he, along-with other Gokhale Institute scholars, has done commendable research. I am indeed 'intrigued' by his withholding action, so to speak. After all, he could rebut me with cost estimates of Maharashtra -- Kannan and Pushpagadan did offer, when criticized by me (Dhawan 1988), the alternative estimates to support their contention of cost margin being against major irrigation works in that state. May be, he has not yet done finely-tuned estimational work in this area, yet has sufficient basis to question others. But, then. I am not offering any summary estimates of two types of irrigation. In point of fact, the main thrust of my paper is somewhat different, namely, to put the cost side of the big dam controversy in right perspective.

I had attempted to wise up the interested reader on three pertinent matters. The first and the foremost was to underscore the long - run compulsions of developing all irrigation sources irrespective of the comparative advantage in cost. The second was to debunk the widely-held (and still more widely propagated) idea that minor irrigation in general and groundwater irrigation in particular is a *low-cost proposition*, or its development cost is a small fraction (*ranging from one-fourth to one-tenth*) of the corresponding development cost for major irrigation works. The third was to draw the prospective researchers' attention to the *inherent limitations and other pitfalls* of working out unit cost estimates for major and minor irrigation works from our data sources, especially from those given in our Five Year Plan documents. I am glad that Mitra has attempted to contribute his bit in this third area. Certainly, there should be no double counting of costs, nor any omission of relevant costs of major irrigation works.

* They report unit investment cost of about Rs. 32,000 per ha for minor irrigation and Rs. 67,000 per ha for major irrigation -- when both are set up in the public sector. Readers may note these are historical costs.

To what extent future empirical investigations will falsify my assertions in favour of major irrigation has yet to be seen. Even if these are proved false, no serious policy changes in water planning would arise. But I have no vested interest, researchwise or otherwise, in promoting the cause of big dams. If I appear either a 'crusader of big dams' or to be 'betraying the symptoms of a lobbyist for the cause of big dams', it could be due to one or both of the following two reasons.

1. lack of communications on my part ;
2. error of identification often committed by parties to a dispute which see in black-white colours, in the sense that if "you are not with us, you are against us."

I have done no project funded research for any body, including Central Water Commission, that apex body for major irrigation development in the country. My research at the Institute has been supported by the Indian Council of Social Science Research (ICSSR) – through a general research grant to the Institute – and by the Ministry of Agriculture since the mid-eighties. Earlier, the responsibility for minor irrigation development was with this Ministry, but now it is with the Ministry of Water Resources. I hold no brief either for the Yojana Bhavan, from which I have kept well-measured distance, notwithstanding my Institute enjoying a good rapport with it. Whatever little hobnobbing with officials/experts has occurred, it has been mainly with those connected with groundwater which has been my main area of research. If at all I hold any brief, it is on behalf of truth that is more often both harsh and bitter than sweet or romantic. Without meaning any disparagement to Dr. Mitra, truths can appear 'strange' or 'shocking' to those professionals who have been long on advice but short on actual truth-seeking. The scope for shocks is ample in Indian conditions if scholars evince greater inclination towards oral discussion than probing into original documents and primary data sources.

REFERENCES

Dhawan, B.D. Irrigation in Kerala-Discussion, *Economic and Political Weekly*, December 10, 1988.

Kannan, K.P. and Pushpagadan, K. Agricultural Stagnation and Irrigation in Kerala, *Economic and Political Weekly*, May 13, 1989, p. 1067-68.

8

Point, Counterpoint

Manu Shroff

It is unfortunate that the Narmada project, which went through a chequered history of inter-state dispute for more than two decades, is now under attack by the environmentalists. "Unfortunate", because all these years they did not raise serious objections. All that was in dispute was the sharing of benefits (and costs) among the states rather than the rationale for a large multi-purpose dam. That the "social awakening" has come late is only one aspect of the problem. That it is possibly influenced by the world-wide concerns for environment is another.

Neither, however, would justify ignoring the critics, for there is little evidence to show that they are inspired by anything other than public interest. In this brief review, therefore, no account will be taken of the charges and counter charges about some of the environmental lobbies acting at the behest of foreign agencies or being funded by such agencies. Nor will any analysis be offered of the differing positions of the various activist groups.

The discussion will be in general terms in an effort to focus attention on what may be called the central issues. These are : the protection of forest cover and resettlement of the so-called oustees. There is no presumption that these things *cannot* be done or that they are *unlikely* to be done, given the "class character" of our society !

But first of all, why a big dam? In the mid-sixties, the green revolution was spurred, among other things, by the massive exploitation of the country's ground water resources. Tubewells, being a cultivator-centered private investment, cut through the deficiencies of large-scale public projects and the pay-off has been good. *One does not need the environmental argument to demonstrate the benefits of wells or minor irrigation in general.* But experts would agree that canal irrigation and the optimal exploitation of ground water have to go together. They are often *complementary* rather than *competitive.*

Also, in the Gujarat context, the critics have not established the feasibility of ground water alternative, especially in the coastal areas which are necessarily afflicted by salinity. The big debate about large dams versus small dams may have a new environmental content : but this should not detract from the "conventional" arguments for minor irrigation in conjunction with major schemes which clearly remain valid.

Indeed some of the alternatives that are offered (e.g. extensive use of lift irrigation in, say, Panchmahals where the Narmada canals will not reach) are not alternatives at all. They are complementary to canal irrigation and should be pursued simultaneously. The Narmada Tribunal itself had gone into the small dam option before giving its decision.

Environmentally speaking, the most important factor in the big-small controversy is the extent of submergence. A large dam shows up as submerging large areas. A series of small dams do not. A cardinal technical fact is that for a given storage volume, a much larger area gets submerged when the storage is obtained through a number of small dams than by a large one. This is a telling argument but it is not conclusive. For, the submergence of a large contiguous area in an ecosystem can create an environmental impact that is proportionately much bigger than that resulting from a number of small dams spread over a large area.

Thus the project authorities have to show that they are doing all that is necessary to "take care" of the adverse environmental consequences and the uprooted people. That this is feasible is itself denied by the activists. But their arguments do not carry conviction. If governments are inefficient, they will be so, whether they are engaged in implementing large dams or a number of small ones.

The arguments which rest on government's inability (leave alone the willingness) to implement the so-called soft parts of the project which involve afforestation and rehabilitation of the "oustees"- the two major concerns of the activists - imply, therefore, that the activists are against government-sponsored and government-directed development.(Indeed, many activists bring in their other biases in the argument against the project, which are wholly irrelevant to the environmental impact-e.g. large contracts, benefits to multinationals etc.)

On the twin questions of afforestation and resettlement, the facts are laid out by the spokesmen of the Narmada Nigam. There is some substance in the activists' claim that on the crucial question and allotment of land to the "oustees" there has been considerable delay. Asking the "oustees" to locate the land and bring the party to the authorities for the negotiation of sale is totally impractical. Who will listen to the poor "oustee" knocking at the door of a rich farmer for a piece of land? But there is evidence that the authorities have learnt by experience and that there is faster progress now, partly in response to the agitation as also to the concerns expressed by the World Bank.

A satisfactory solution to the human resettlement problem is, without doubt, the key to the success of the project since it is, perhaps, for the first time that the project itself contains provision for it. The recent indication that the World Bank may be willing to finance even the acquisition of land for resettlement should make for faster progress.

Submergence of centuries old rich forest is a major concern. The destruction of the flora and fauna, rare species and, more important, a forest that is a naturally well balanced organic whole (and hence not replicable by human effort) cannot clearly be dismissed as irrelevant. Studies are afoot to assess the damage and to find out to what extent it can be repaired. Not enough is publicly known about this aspect of the project. But it should not be mixed up with the loss of trees as such--since that can surely be made good by growing them elsewhere?

Environmentalists lack clarity in their presentation when they put out figures of "cost" by aggregating the cost of land and timber both of which are scarce but can be found elsewhere. (Indeed land is "replaceable" by increasing its productivity, which is precisely the objective of the project). The critics should confine their cost calculations (if at all they are feasible) to the irrevocable loss due to submergence. This has not been assessed.

But to mix up this genuine concern with another, namely preservation of the Life-styles of the inhabitants is romantic nonsense. Let those who are affected be pooled on *life-styles.* They are more likely than not to choose a higher standard of life even if it meant a change in their traditional style of living. In the name of protecting adivasi culture we are in the danger of creating "reservations" which should be anathema to a democratic polity.

Finally, the economic arguments. It is not for the envrironmentalists to worry about the economics of the project. In doing so they tend to weaken the "greenness" of their case. Anyone who knows the World Bank's bureaucracy should know that that agency would not touch a project if the rate of return was not adequate or acceptable.

In the case of the Sardar Sarovar Project, which the Bank is financing, the internal rate of return has been calculated at 13 per cent and the sensitivity analysis shows that on various assumptions about variations in costs and benefits, the range widens to 7 to 19 per cent with the expected rate of return being 12 per cent. Infrastructure projects are known to have relatively low rates of return (one reason why they have to be in the public sector).

The return is close to the opportunity cost of capital. True, costs will go up by the time the project is completed-mostly due to inflation. But so will the value of benefits. No one can predict how relative prices will change. Also, it is only at the initial stage that the cost-benefit analysis is usually done. No revision is made, unless a *major* disturbance to relative prices (like the oil crisis in 1970's) occurs.

Some calculations have been made by the critics of the project (e.g. the authors of the INTACH report) on the assumptions that the base cost goes up by 19.5 per cent, there is a two-year delay in implementation *and* the utilisation of irrigation potential goes down by 25 per cent. They show that even on these assumptions, the economic rate of return works out to 10.8 per cent. That this is lower than the rate calculated by the TECS or the World Bank is beside the point. That it is still not lower than 10 per cent (in an infrastructure project) is of great significance.

Another economic argument against the project is that it involves large resources which the government does not have. A whole lot of irrelevant noises are being made in this context, including charges of misleading the public. The short point, however, is that the annual burden, at least in the case of Gujarat, is not too heavy. Also, the alternative of a series of small dams (with the same benefits) is unlikely to cost less.

But then, what kind of development will the project create? In the case of Gujarat, the argument being made is that the rich will benefit, not the poor. The former is true. The latter is not. Per capita income in the drought-prone areas are notoriously low. And 75 per cent of the command area is stated to be drought-prone. The average level of incomes in these areas will go up as a result of the project. The critics have to demonstrate that, alongside, the inequality will also go up-more than at present. This they have not done.

At bottom, the conflict between development and protection of environment is more apparent than real. Clearly, the very activists who complain about de-forestation are often ranged on the side of the *customary* right of the tribals to "enter" the forest to collect fallen branches and twigs (also, since they have no alternative fuel available). They connive at the illegal felling which occurs once the right of entry is recognised.

Poverty is the worst enemy of environment. Its alleviation requires, to paraphrase the Bruntland Commission report, "human intervention with natural systems during the course of development." It is futile to complain about all such interventions and hanker after conservation for its own sake. Settled agriculture was perhaps the most massive intervention with the eco-systems. But the latter was not destroyed.

Nature's regenerative powers have been immense. It is when the intervention is both massive and threatens "life-support systems *both globally and locally (emphasis added)* that the problem arises. The Bruntland Commission hit the nail on the head when it said : "At a minimum, sustainable development must not endanger the natural systems that support life on Earth : the atmosphere, the waters, the soils and the living beings". The activists have notshownthat the Narmada Project has "endangered" any of these beyond repair.

9

The Sardar Sarovar Controversy
Are the critics right?

Rusi Engineer

The fierce controversy currently raging over the Sardar Sarovar Project--and the growing public opposition to large dams and other development projects--is a phenomenon so unfamiliar and perplexing that the commonest response to it is still one of blank incomprehension. At the same time, the anti-dam movement is steadily gaining strength and popular support, and now includes within its ranks some of the most eminent and distinguished individuals in this country.

One of them, Baba Amte, has put his life on the line and publicly sworn a solemn oath : "While there is breath in my nostrils. I will not allow this dam to come up." He has issued a passionate challenge to the authorities : they can go ahead and fill the dam if they wish, but they will have to drown him first.

And when that time comes, Baba Amte will not be alone; thousands will be with him in the submergence zone. A whole new movement has emerged against "destructive development", consisting of a wide assortment of social activists, intellectuals, environmental and human rights groups, with increasing grassroots support among project-affected people in several different parts of the country.

At the same time, it remains true that most people, including a majority of our economists and development planners, simply do not see what the activists and critics are driving at and what the fuss is all about. They tend to dismiss the antidam activists and critics as sincere but misguided people, as latter day Luddites who are afraid of the social dislocation that inevitably accompanies growth and development, backward looking romantics who wish to put the clock back and stop the march of progress.

The incomprehension that has typically characterised the controversy over the Sardar Sarovar Project is in evidence each time an attempt is made to answer the critics. It is blithely asserted that the critics have not made their case--when it can be readily shown that the real issues that arise in connection with the project have not even been correctly perceived, let alone adequately addressed.

In what follows, we shall not let ourselves be distracted by the plethora of facts and figures put forward by the government and disputed by the critics, nor dwell on the welter of claims and counterclaims, accusations and recriminations, that have obscured the deeper issues. Many of these points of dispute--such as the gross underestimation of the siltation rate and the consequent overestimation of the life-span of the reservoir, major lacunae in the planning process which will inevitably reduce the projected benefits, manipulation of the cost benefit analysis with deliberate underestimation of costs and overestimation of benefits which place the viability and the project in doubt, denial of the public's Right to Know--are indeed of considerable importance. But in the present situation, it is even more important to grasp the essence of the phenomenon and not miss the wood for the trees.

CENTRAL ISSUES

Two sets of issues are central to the Sardar Sarovar controversy: first, resettlement and rehabilitation of the nearly 100,000 oustees (nearly 230,000 if we also take Narmada Sagar into account), and, second, the ecological implications of the project.

Taken together, these two issues have far-reaching implications that carry us well beyond the pros and cons of a particular project: on the one hand, we are led to a critique of the prevailing model of development which is increasingly seen as socially unjust and ecologically unsustainable; and on the other hand, we are brought face to face with the exciting prospect of an alternative model of development, based on ecological restoration prospect of an alternative model of development, based on ecological restoration and sustainable livelihoods for all our people.

Let us first look at the problem of resettlement and rehabilitation (R&R). The establishment view does not deny that there has been considerable delay on the crucial question of acquisition and allotment of land for the oustees. But it takes a lenient view of the delay, preferring to believe, in the words of *The Economic Times* editor, that "there is evidence that the authorities have learnt by experience and there is faster progress now, partly in response to the (oustees') agitation as also the concerns expressed by the World Bank". In other words the problem is conceded, but we are asked to believe that it is being taken care of now. Having thus satisfied their consciences, and got this inconvenient little matter out of the way, they are ready to get on with the project.

The activists working with the oustees' organisations, on the other hand, have not been able to find any such "evidence". Having had the bitter experience of dealing with the government of Gujarat on the oustees' behalf for the last five years, they can see that the oustees' interests are being sold down the river. It is clear to them that the project authorities are least concerned about the tedious and bothersome business of R&R since the big bucks and the best career prospects are elsewhere. In any case, the oustees have little choice but to curse their fate and accept whatever treatment is meted out to them. To the activists, however, the fate of the oustees has become a major human rights issues, involving the fundamental right to life and livelihood of the poorest and weakest sections of our people.

VOICES OF PROTEST

To see the issue a little more clearly, let us for a moment put aside intellectual disputations and listen to one of the many authentic voices of the new activism. Here is a newspaper report of Srilata Swaminadhan addressing a gathering of some 35,000 project-affected tribals and small farmers last month at the National Rally against Destructive Development in the little town of Harsud in Madhya Pradesh :

" 'I have come,' says this fiery activist from Rajasthan who was physically attacked last year by people who took amiss her work on behalf of the tribals, 'all the way from Rajasthan to tell you something. They built dams in Rajasthan, too. There too they promised all sorts of things–land for land, compensation in the thousands, everything. The people trustingly gave up everything they had. Do you know where they are now? Scrounging around in the garbage heaps of big cities, hoping to snatch a morsel of food for their hungry bellies--believe me, I work with these people. That's why I have come to tell you, don't believe this government. Don't even listen to their promises. Above all, don't allow this dam (Narmada Sagar) to come up and flood you out of your hearths and homes. Tell me, will this dam be built? *Nahin,* comes the answering roar, but Srilata is unhappy : 'Unless you shout louder they can't hear you in Delhi,' she says. *Kabi nahin,* the crowd yells back, and this time the echoes shake the earth."

Let no one mistake this for rabblerousing; it is the authentic voice of protest against injustice. Still, for those who would prefer a more sober tone, here is the May 1989 report of Thayer Scudder, consultant to the R&R sub-group of the World Bank's Reappraisal Mission for the Sardar Sarovar Project that visited India earlier this year. Incidentally, Scudder is professor of anthropology at the prestigious California Institute of Technology and one of the world's leading experts on R&R.

In the first sentence of his report, Thayer blandly announces that while R&R in connection with large-scale projects is always difficult, it is especially so in the case of Sardar Sarovar "since there is little precedence in India for execution of equitable R&R in connection with development projects". Translated into ordinary language, Scudder is saying that the treatment of oustees in this country has always been less than fair, that they have been uprooted and more or less left to fend for themselves, and consequently, as Srilata Swaminadhan asserted, they have almost invariably been pauperised and reduced to destitution.

SHOCKING NEGLECT

Scudder's indignation at what he found comes across even through the bureaucratice prose of his report, which is a scathing indictment of the R&R aspect of the Sardar Sarovar Project : "Ten years after the Tribunal rendered its decision, poor implementation continues to be the main deficiency of the R&R component... It is important to examine not only what has been avvomplished but also the'sprit' in which such accomplishment has occurred... The favourable oustees, policies of Gujarat have resulted primarily from external pressure rather than from the 'good will' of the Government of Gujarat (GOG) towards its oustee population.

"As documented in detail in this report, the GOG has attempted neither to implement expeditiously its new policies nor to correct defiencies (in the resettlement) of these relocated prior to December 1987 (when the new liberal policies began to be announced). During April 1985 I made a special effort to search out examples of 'good will' on the part of GOG towards its oustees in regard to a number of long-standing trouble points which the government at small financial cost, and at significant benefit to oustees, could easily have achieved. *I found no such cases.*"

So much for the boast often repeated by the project authorities, that the compensation norms established by the Gujarat government are "the most liberal not only in India but perhaps the world". For, what use are liberal norms when the government shows every day by its actions that it has no intention of implementing them?

It is important to note that the activists and oustees' organisations which have now taken a firm "no dam" position and are adopting an increasingly adversarial, even militant, stance, have been driven to this position only relatively recently. To begin with, they came into the picture with the intention of merely acting as intermediaries between the government and the displaced villagers, to help the oustees to actually get what was being promised to them by the government.

Scudder's report provides ample corroboration of the fact that it was only when the oustees' organisations discovered, after years of bitter experience, that the government was not sincere in its promises and was trying to cheat them, that their stance finally hardened into the un-compromising position: *'Koi nahin hatega, bandh nahin banega'.* If Scudder after making a "special effort to search out examples of goodwill on the part of GOG", could find "no such cases", the oustees can hardly be blamed for resisting the government's efforts to uproot them from their homes and drive them from their ancestral lands.

The demand of the oustees' organisations is simple, unexceptionable, and perfectly just. All they are saying to the government is: if you want us to give up our homes, abandon all we have and move out lock, stock and barrel, at least give us some tangible proof that your promises are sincere and that you mean what you say. Show us, on paper, what land you have acquired for us; show us what plans you've formulated as to where and how we are to be rehabilitated, what arrangements you've made to see to it that our relocation to some unfamiliar far-away place will be accomplished reasonably smoothly and without undue hardship. Above all, show us how you've actually rehabilitated the handful of villages displaced since 1970-71, and let that example convince us that it will be to our advantage to move. If you can't do any of these, you have no right to ask us to move.

SEVERELY CRITICISED

The resettlement of the few displaced villages already carried out-which was to have been a model that would persuade the rest of the oustees to move--has in fact been severely criticised not only by the World Bank but also by the official monitoring agencies appointed by the government itself, viz, the Tata Institute of Social Sciences (TISS), Bombay, and the Centre for Social Studies (CSS), Surat.

Scudder relates in his report that whenever the CSS pointed out what they considered to be planning or implementation weaknesses, GOG replied that their conclusions were invalid since everything was on schedule. And he goes on to add, "In fact, everything is not on schedule, in part because of the total absence of date specific targets relating to all aspects of land identification, acquisition, purchase, titling and development process."

If further proof is needed, the National Institute of Construction Management and Research (NICMAR) -- the official agency appointed by the Narmada Control Authority set up by the Central Government to monitor compliance with stipulations laid down in the Narmada Water Disputes Tribunal award and in the World Bank loan agreement--states in its January 1989 report: "As a Central monitoring and evaluation agency, we tried to locate if there was any detailed implementation plan for the R&R work. We have not come across any such plan, Central or interstate or even at the level of each involved state."

To cap it, in July this year, following the adverse report of its reappraisal Mission, the World Bank has given the concerned state governments one more year to come up with acceptable policies and plans for R&R; and if they fail to do so by 30 June 1990, the Bank has threatened to cut off its credit to the Sardar Sarovar Project. (That the threat comes too late, because by that time the project may be irreversible and the oustees' fate would have thus been sealed, is another story).

All this is bad enough, but worse is to come. The problem is not only in the implementation. The fairly liberal R&R policies announced by Gujarat apply only to less than 20 per cent of the oustees who are to be resettled in Gujarat. A large majority of the oustees - more than 80 per cent - are from Madhya Pradesh (M.P.) and Maharashtra and are to be resettled in those states. Gujarat's liberal policies will not apply to them: they will be dealt with under the R&R policies of their own states which are much worse in decisive respects. Yet the large newspaper advertisements issued by the project authorities continue to mislead the public into thinking that Gujarat's liberal policies are applicable to all oustees.

BETRAYAL OF OUSTEES

Thus it is clear that as of today there is nothing even remotely resembling a comprehensive and detailed action plan for R&R, and that acceptable policies applicable to all oustees do not exist. Nor have the project authorities bothered to set up a strong, functional agency for carrying out R&R, staffed with people having the necessary skills, dedication and compassion, without which there is no chance of successfully accomplishing the difficult and enormously complicated business of rehabilitating such a large number of people. Nor can anyone realistically claim that the requisite policies, plans and implementing machinery will be brought into being in the short time before the remaining displacement is due to begin.

The conclusion is therefore inescapable: if construction of the dam is allowed to proceed, there is little doubt that a large number of poor people will be pauperised and condemned to a lifetime of destitution. A grim fate awaits a large proportion of then oustee population -- futile wandering in search of work, toiling in the most exploitative conditions, begging on city streets and scrounging around in rubbish dumps, shunned by all, and in the end reduced to such dire straits as to be forced to sell their wives and daughters. Such is the sacrifice being demanded at the altar of "progress".

Such a demand is unacceptable, not only to the oustees themselves, but to a large and growing cross-section of concerned citizens who believe that under these circumstances the project must not be allowed to go on. As Baba Amte has aptly said, we simply cannot stand by and watch some people's "life-line" being allowed to become other people's death line.

It is necessary to remind ourselves that equitable R&R is not a matter of state generosity: it is a fundamental right of the oustees guaranteed by the Constitution. The concerned state governments on their part are legally bound to rehabilitate all oustees in accordance with the norms and stipulations laid down by the Tribunal and the World Bank-- not to mention their own policy pronouncements.

VIOLATION OF RIGHTS

The appalling lack of preparedness with regard to R&R is so well documented that the government's wilful breach of these stipulations, amounting to a violation of the oustees' fundamental rights, would be easy to establish in a court of law. In such circumstances, any court- provided it does not have one law for the powerful and another for the weak-- would be constrained to issue a stay order against the project until the authorities have demonstrated their compliance with stipulations.

The argument holds irrespective of the magnitude of benefits expected from the project; indeed, one could say that the greater the magnitude of benefits, the greater of obligation of the government and the beneficiaries to adequately look after those whose sacrifices have made those benefits possible.

So far, it has been argued that the government *has not* made acceptable arrangements for adequately resettling the oustee population; but it is possible to go further and show that such arrangements *cannot* be made in the foreseeable future. Given that R&R is to be carried out on a 'land for land' basis, the crux of the problem is the non-availability of suitable land. Unutilised cultivable land outside forest areas simply does not exist, and the Central government has been rightly refusing to release forest land for resettlement purposes. The only remaining possibility is to resettle oustees on private irrigated land purchased by the government for this purpose in the command area.

While this option is technically feasible, it is known to be politically unworkable, for it involves compulsory acquisition of irrigated land from relatively rich farmers (only they have any surplus that could be acquired) at well below market prices, and handing it over to oustee families, which almost always belong to social strata much lower than the host community in terms of caste, social status and power. Their successful rehabilitation under these circumstances is virtually impossible.

As the experience of legislation in Maharashtra (1976) and (M.P.) (1985) enabling state governments to compulsorily acquire land in the command area has clearly shown, state governments lack the political will to enforce such legislation, which merely remains on paper. (The same has been the fate of land reforms legislation which has remained largely unimplemented, and for the same reasons.) Thus if we consider -- as we must - the social and political context in which R&R has to be carried out, the conclusion cannot be avoided that satisfactory resettlement and rehabilitation of the oustees is just not possible.

WHO GAINS? WHO LOSES?

This is an astonishing conclusion, for it means that, in the present socio-political set up, grave injustice towards the poorest and weakest sections of our people is an intrinsic feature of all development projects involving large-scale displacement of people. We can now begin to understand the significance of the increasing mass mobilisation and public opposition to large development projects -- such as dams, mining projects, thermal and nuclear power stations, and even defence related projects like the missile testing range at Baliapal. All these development projects sharply raise a potent political issue: "Who gains from such development? And who loses?"

The Sardar Sarovar oustees have been told that the project will brook no interference and must go through because it is in the "national interest". the chief minister of Gujarat has publicly declared his willingness to call out troops, if necessary, to deal with any attempt to obstruct the construction work. But the oustees are no longer in a mood to submit meekly to this sort of thing. The spontaneous question to which they are now demanding an answer is: "We are told that we must sacrifice for the nation. But who really comprises our nation? Are we also not part of this nation?"

It is a question that is becoming increasingly insistent. In almost 40 years of so-called planned development, many of the oustee villages have not seen an iota of improvement in their living conditions, not a single ray of progress has penetrated there. They do not have clean drinking water, schools or the most elementary health care facilities; people have little productive work, and employment opportunities are non-existent. How can we ask these people to make sacrifices for others when no one has ever sacrificed anything for them, or even spared a thought for them? The simple truth is that planned development has passed them by. Not only has it done very little for them, it has often left them worse off than before.

POLITICISATION AT GRASSROOTS

The oustees are now refusing to take thing lying down. They are no longer prepared to become victims of other people's development. Their self-assertion is part of a wider politicisation of the many who have been marginalised and excluded from the gains of development. The rising tide of protest against "destructive development" we are witnessing today is the backlash of our lop-sided development policies that have created relatively small islands of affluence amidst the vast surrounding ocean of poverty and deprivation. In agriculture as well as industry, in rural as well as urban areas, the benefits of growth and development have been highly unevenly distributed. Wealth has remained confined to the relatively few and has refused to 'trickle down' fast enough to the many. The oustees' protest draws its strength from the growing unrest at the grassroots which is the inevitable consequence of unjust and unequal development.

The stock reply to this sort of criticism has always been that what we need is not less development but more and faster development. The fact of unequal and lop-sided development is not denied, but it is maintained that growth has not trickled down and spread more widely simply because there hasn't been enough of it.

This brings us to the second major issue we have to consider. The prevailing model of development is under challenge not only from the point of view of equity and social justice, but also because of its disastrous ecological consequences. Thus, more growth -- of the present kind -- would also mean a bigger ecological disaster.

EARLY PHASE

When 'environmentalists' take a stand against Sardar Sarovar or any other large project, what exactly are they saying? It is important to distinguish several distinct strands in their argument. In the early phase of the environmental movement, the emphasis was on the conservationist or preservationist impulse which seeks to prevent the destruction of natural forests and wildlife occurring as a direct consequence of the project. For instance, Sardar Sarovar and Narmada Sagar together will submerge 54,106 hectares of forest and 55,681 hectares of cultivable land.

No one any longer denies that the submergence of a centuries old forests, with its profusion of flora and fauna, including rare species, is an unmitigable tragedy and a major cause for concern. Once destroyed, this intricately balanced community of plant and animal life, evolved over millenia, is impossible to recreate through human effort and for which man-made plantations are no substitute. To this was added a recognition of the crucial role of forests in soil and water conservation and in the regulation of climate, and their value as gene pools and as a storehouse of biological diversity.

The environmentalists are concerned that large projects are helping to destroy the last vestiges of this country's forest cover at a time when the crying need is not only to safeguard all our remaining forests but to substantially *increase* the area under forest. It is true that loss of forest due to such projects accounts for a relatively small proportion of the total deforestation taking place each year. But the destruction in each case sharply brings home the fact that, despite much talk, the government does not know how to even partially make up for the loss by means of "compensatory afforestation", nor does it have a workable reforestation strategy for restoring vegetational cover over vast tracts of this country's barren and degraded lands.

In recent years, however, the environmentalist viewpoint has matured and evolved much beyond the limits of the conservationist perspective. Its focus has broadened and shifted from 'conservation' to 'ecologically sustainable devlopment'. Large dams (and other mega-projects) are now being opposed not only for the direct destruction they cause, but because they are seen as key features of a type of development that is playing havoc with our natural resource base and life-support systems, and is therefore unsustainable.

SUSTAINABLE DEVELOPMENT

The concept of 'ecologically sustainable development' is so new that its far-reaching implications have by no means been fully articulated. Nevertheless, it is a concept that answers a deep need of our times, and provides the conceptual underpinning for a radical critique of the prevailing model of development, as well as a basis for an alternative model of development.

The public animus being directed against Sardar Sarovar (and other large scale development projects) is not due only to the direct damage the project will cause, in terms of environmental destruction and human displacement. Ultimately, what fuels this animus is not just an individual project but the whole pattern of development, of which these projects are an integral part. It is necessary to understand what this pattern of development is, and why it provokes the ecologists' wrath.

The purpose of a giant multipurpose project like Sardar Sarovar is to provide large quantities of irrigation and power. the basic goal is to extend the use of new production technologies that will lead to increased production in both agriculture and industry. In agriculture, the project will help to usher in the Green Revolution over a large area; and the power it generates is essential for industrial development.

These goals appear perfectly laudable and benign, until we realise that there is a snag: the process is destructive and unsustainable. The Green Revolution route to agricultural development locks us into a pattern of resource use that has no future. Every ingredient of this particular technology package – high levels of fertiliser and pesticide use, high yielding varieties of seeds (HYVs), and intensive irrigation– is riddled with problems.

It is a technology that demands high and continuously rising inputs of chemical fertilisers and pesticides to maintain yields, but forgets that these critical inputs are petroleum-derived products, based on a depleting stock of fossil fuels that may be exhausted in a generation. The Green Revolution pattern of agriculture is sustained by oil : what happens when the oil runs out?

In addition, depleting supplies of fossil fuels also mean continuously rising prices of farm inputs which further fuel inflation. Worse, high levels of use of chemical fertilisers result in declining soil fertility over time and the eutrophication of water bodies, while pesticide use creates serious and well known health hazards--toxic contamination of food, high concentrations of DDT in mother's milk, carcinogenic PCBs in human brain tissue. The hazards of manufacture need no reminder – a pesticide plant was responsible for the Bhopal tragedy.

The use of HYV seeds has its own problems, ranging from the risks of monoculture, depleting gene pools of plant cultivars, dependence on a small number of artificial varieties of seeds and their built-in susceptibility to new pests and diseases, due partly to the phenomenon of pesticide resistance among pests. All these things, taken together, leave little room for doubt that a system of agriculture based on such foundations cannot long endure.

Already, in the US--the home of chemicalised agriculture--serious rethinking is under way and has already achieved official recognition. The prestigious National Academy of Sciences as well as the US Department of Agriculture have recently acknowledged that natural farming using modern techniques is now a fully viable option, and have recommended policy changes that would help transform US agriculture in that direction.

RESOURCE ILLITERACY

The third ingredient of the Green Revolution package--intensive irrigation--presents a different kind of problem. Our irrigation engineers and planners believe that the best solution to the acute water problem of our drought-prone, semi-arid areas lies in large-scale transfers of surface water over long distances through large irrigation projects. B.B. Vohra, a lone official voice crying in the wilderness, has aptly characterised such thinking as a product of the "resource illiteracy" or ecological ignorance of our planners. Sardar Sarovar provides a prime example of such thinking in–as-much as the solution to the acute water crisis in Kutch and Saurashtra is sought in the waters of the Narmada transferred, at enormous cost, over long distances.

MAN-MADE PROBLEM

To grasp this matter a little more clearly, we need only to recall that many of our so-called drought-prone areas were not always so--that their present desiccated condition and acute water scarcity is not natural but man-made. Till fairly recently, say 50 to 70 years ago, Saurashtra was a green and prosperous region, and its rives, tanks and wells had plentiful water all the year round. There is historical evidence to show that several regions which are today barren, drought-affected and almost desert like, were once covered with rich pastures or dense forests.

For instance in the 18th century, when Sarbuland Khan, the regional ruler of Gujarat, attacked the capital city of the Jethva rulers of Saurashtra, his cavalry could not advance due to dense, impenetrable forests, and could reach its destination only by cutting and burning its way through it! Similarly, the barren and chronically drought-affected Jhabua district in MP was thickly forested within living memory.

It is destruction of the once rich pastures and forests of these areas which is the real cause of their acute water crisis today. It is crucial to realise that the low rainfall in these areas is not the main culprit, as can be seen from the fact that, despite low rainfall, wells used to have water all the year round. Rivers which are today drying up a month after the monsoon used to flow almost perennially when the land had not yet lost its vegetational cover. Conversely, the connection between water scarcity and loss of vegetation is clearly seen in high rainfall areas in Kerala or the Himalayas where acute water shortages are developing in the wake of deforestation.

Once this connection is seen, restoring vegetational cover to the land becomes the first priority. In B.B. Vohra's memorable phrase, a massive "internal haemorrhage" has been going on in the country's natural resource foundations which is bleeding the patient white. In this situation, to run after complex and enormously expensive technological 'solutions' like the Sardar Sarovar Project is to be like the man who refuses to look after his body sensibly and continues to inflict on it every kind of abuse--heavy smoking and drinking, foolish eating habits, no exercise, constant stress--and then looks to the marvels of medical science to fix any problems that arise.

The analogy is pretty exact. Just as sensible care of the body effectively prevents many ailments, but does not mean that hi-tech medical intervention will never be necessary, so also proper care of the land could effectively eliminate many a water problem, without implying the *all* problems could be solved in this way and that large dams will never be necessary. As with the body, so with the land. Sensible and inexpensive steps to maintain health ensures that costly and sophisticated technological interventions will be needed less often, and in some cases avoided altogether.

This is not romantic and fanciful 'back to nature' nonsense. Village level efforts in several drought-prone areas of the country have conclusively demonstrated the truly amazing results that are possible with a programme of small watershed development and ecological restoration, rainwater harvesting and small-scale storage, and increased efficiency of water utilisation.

The village of Ralegan Shindi, in the severely drought-prone Ahmednagar district of Maharashtra, is one well-known example among many. Rainfall here is low and highly erratic varying between a mere 200 mm and 800 mm, with an average of 574 mm--not very different from large parts of Gujarat to be served by the Sardar Sarovar Project. Before the watershed development programme was initiated just a few years ago, the village faced acute drinking water shortage, yields of jower and bajra were very low, and more than half the food requirements of the village had to be purchased from outside. After the watershed programme, drinking water shortage has been totally eliminated, irrigated area (in terms of limited water application) increased twelvefold, the cropped area as well as yields have more than doubled, and the village is now a net exporter of foodgrains.

EQUITABLE AND SUSTAINABLE

All this has been achieved in just a few years, with limited means, without using any exogenous water from any large system--and results are expected to be even better in the coming years. This, and several other similar examples have clearly shown that with this approach it is possible to meet all reasonable water requirements in rural areas, without the human and environmental costs invariably associated with large projects.

It is important to note that this approach is inherently equitable, in-as-much as it needs to be implemented over the whole of the country's cultivable area. In contrast major or medium irrigation projects are location specific, and represent a concentration of investment capable of benefiting only a small fraction of the country's cultivable area.

If the colossal sum to be spent on the Sardar Sarovar Project--the latest estimate is Rs. 13,500 crores--were to be spent on implementing the new approach, it could achieve results like those in Ralegan Shindi in every village in Gujarat. What is wrong with large projects like Sardar Sarovar is well expressed by Baba Amte : "There must be sufficiency for all, before there is superfluity for some." In the new dispensation, recourse would be had to dams, of the smallest size necessary and mainly to meet urban needs, only to the extent that a deficit still remains after fully implementing the new approach.

So, it turns out that the ecological viewpoint shows us a whole new way of thinking about development--the kind of development in which the need for ecological sanity and the demand for social justice coincide and mutually reinforce one another. This convergence of ecology and equity is perhaps the most promising and creative new development in the country today.

It is not too much to claim that we are witnessing the emergence of a new developmental pradigm, a new way of seeing--of which the critics of 'destructive development' and the activists fighting against large projects are the early representatives. Admittedly, in view of its recent birth, the new paradigm is as yet only imperfectly articulated and all its ramifications have not been clearly spelt out. But the power of its central vision is already evident and its coherence is growing with every year that passes.

FALSE TRAIL

According to the new vision, our economists and planners have been on a false trail, Instead of focusing on people and on how to improve their lives, they have sought to increase production and maximise GNP. They have tended to forget that the ultimate objective of planning is not to produce so many million hectares of irrigation or so many thousand megawatts of power. These things are not ends in themselves—they are useful only in so far as they help in eliminating the poverty of vast masses of our people. This the existing model of development has conspicuously failed to do.

SARDAR SAROVAR AND ITS COMMAND AREA

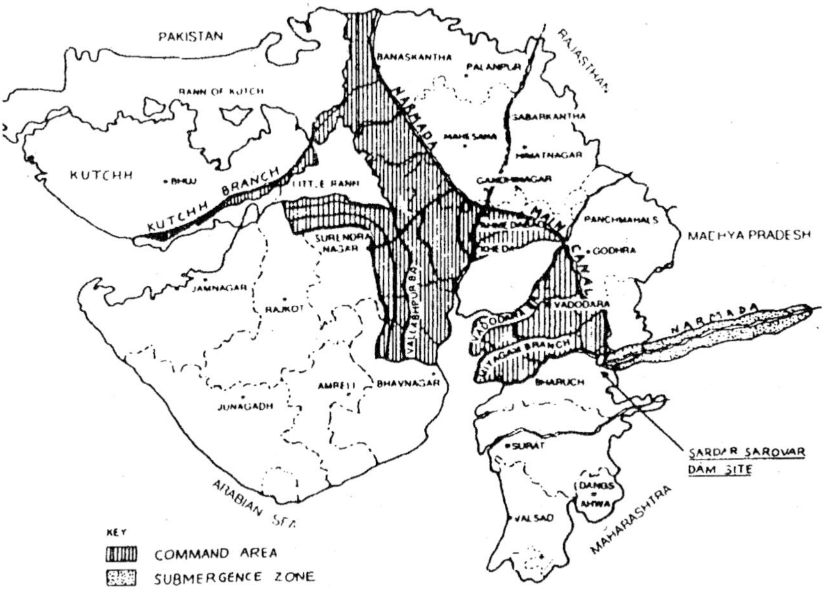

Stupendous cost, unequal distribution of benefits : the most drought-prone and water-scarce parts of Gujarat are outside the command area.

After 40 years of the sort of 'progress' we have had, the gains of development remain confined to the relatively few, and the proportion of people below the poverty line remains more or less unchanged. India continues to be the home of a majority of the world's poor, hungry and malnourished; our so-called planned development has left us poorer than subsaharan Africa; and our per capita income continues to rank near the bottom of the list of the world's nations. Clearly, the old medicine has not worked, and it is time to change the prescription.

A development plan which has the objective of finding decent and sustainable livelihoods for the entire population would look very different from present day plans for so-called development. It would, first of all, put unemployed rural labour to work on ecological restoration and regenerating the natural resource base which is the foundation of all real development. Village by village plans would have to be worked out-requiring substantial decentralisation of planning and decision-making--for small watershed development, afforestation of all barren and degraded lands, soil and water conservation works, rainwater harvesting and small-scale storage, fodder and pasture development, the development of biomass-based energy sources, and small-scale industries based on agri-silvi-pastoral development.

GREENING OF INDIA

A country-wide programme of this sort, for the regeneration of the "uncultivated half of India", urgently needs to be recognised for what it is--a prime developmental activity, fully comparable in scope and importance to agriculture, and forming the core of a new strategy of sustainable development capable of eliminating the worst forms of rural poverty and unemployment within a decade. What makes such a strategy even more attractive is that it presupposes the active involvement and participation of the people, and therefore provides a unique opportunity for a major thrust towards democratisation of the grassroots.

The unprecedented turmoil in Indian society today is not unrelated to the imbalances and injustices inherent in the present model of development. In particular, the resurgence of ethnic self-assertion and widespread tribal unrest is frequently the result of patterns of resource use resembling colonial exploitation.

The Narmada Valley Development Project, for instance, is not a project for the development of the valley's resources for the welfare of the valley's inhabitants. Rather, it is a project for the appropriation of the vallye's resources by outsiders for their own benefit. Irrigation and power from the project will benefit one set of people outside the valley, while the costs will be borne by another set of people for whom the valley has been for centuries their ancestral home.

VIOLENCE AND EXPLOITATION

Such conflicts over the control and utilisation of natural resources are erupting all over the country. In the case of the Narmada valley, the resources in question are the river itself and the surrounding forests and cultivated lands, on which depends the livelihood of hundreds of thousands of tribals and peasants from the backward and weaker sections of our society. On the other hand, the commercially, economically and politically dominant classes from outside the valley want these resources for themselves and are pushing for the project.

It is an unequal contest. The terms of exchange are grossly unjust, just as they were under colonial exploitation. The colonisers possessed military might but lacked the resources; while the colonies had the resources but lacked the power to defend themselves and their interests. It ended with the growth of nationalist consciousness and struggles for independence.

Something very similar is going on in our own resource-rich but otherwise backward, tribal areas. The more developed and powerful parts of the country are dominating and exploiting the more backward and weaker parts. Once again, the unjust and exploitative interaction between the two is creating conditions for the emergence of separatist movements. Where ethnic differences between exploiter and exploited are also present, we have fertile ground for sub-nationalistic aspirations; and once this happens, it is very easy for things to take a violent turn, as we have seen in case after case. The Jharkhand movement is a classic example.

In the Narmada valley, there is already a strong protest movement with considerable mass support. The present leadership is committed to democratic and peaceful forms of resistance; but if a Subhash Gheising were to arise on the scene, the confrontation could easily take a more ugly turn. Violence and terrorism are never far away in such situations. The price of not heeding the activists and critics could turn out to be heavier than anyone imagined.

GREEN POLITICS

In contrast, the alternative development model--ecological restoration and sustainable livelihoods for all our people-offers the only strategy capable of spreading development more evenly and equitably among all classes and to all regions of the country. Coupled with decentralisation and democratisation at the grassroots, it constitutes a vision of ecological sanity and social justice that is extraordinarily powerful, having the potential to develop into a full-fledged political alternative.

Some such vision of a new Green politics-embryonic, dimly perceived and inadequately expressed--is what ultimately drives the activists and critics of Sardar Sarovar and other destructive mega-projects. To ignore them, or to quibble with them because their vision is not yet fully developed and lacks detail, would thus be foolish indeed.

10

Water Resources of India And Its Development Prospects*

K.C. Thomas

India is vast country with a geographical area of 328 Million Hectares (M.ha). It receives an average annual rainfall of 1170 millimetres (mm) which is the highest in the world among countries of comparable size.

India receives most its rainfall from the South-West monsoon originating in the Indian ocean and having two distinct branches, the Arabian sea branch and the Bay of Bengal branch. The Arabian sea branch produces rainfall in peninsular India and part of Gujarat and Rajasthan. The rest of India receives rainfall from the Bay of Bengal branch. The windward sides of the hills and mountain ranges receive high rainfall while the leeward side and the interior of the vast plains receive less rainfall. For example, the Khasi and Jaintia hill areas in the north-east of the country receive as much as 10,000 mm of rainfall whereas western Rajasthan receives only 150 to 200 mm. About 75 per cent of the rainfall takes place in the four monsoon months of June to September. Another 10 to 11 per cent each occur in the pre-monsoon and post-monsoon months of March to May and October to December respectively, the winter rainfall (January-February) being only of the order of 4 to 5 per cent. There is thus a large variation in the incidence of rainfall in different parts of the country and in addition, large variations exist between different seasons. There is also large variations in the rainfall from year to year, usually in cycles of wet years followed by dry years.

---

* This is the First Kanwar Sain Memorial Lecture held in Delhi on 25th January, 1990. The author, at the request of the editor, has expanded the text that was circulated at the time of the lecture.

There have been varying estimates about the total precipitation in the country and its sub-distribution into evaporation and transpiration, surface flow, sub-surface flow and regeneration and contribution to ground water recharge. I am quoting here the comprehensive set of figures assessed by the National Agricultural Commission in 1976. According to them, the average annual precipitation over the whole country, (including snow fall which constitute only a small part, about 2 to 3 per cent) is 400 million hectare-metres (M.ha.m), of which about 70 M.ha.m, evaporate immediately from the top soil, 215 M.ha.m infiltrate into the ground out of which 165 M.ha.m go back to the atmosphere as evaporation and transpiration, and 50 M.ha.m go down to recharge the ground water. It is also envisaged that 5 to 10 M.ha.m will go from stream flows to ground water during floods and further that about 50 per cent of the flow-irrigation water (25 M.ha.m) will also go to ground water. At the same time, 45 M.ha.m of ground water is estimated to re-appear as surface flow (regeneration) in streams and rivers during the low-flow season. The resultant surface flow including 20 M.ha.m received from adjoining countries is estimated to be 185 M.ha.m and the net ground water recharge to be 40 M.ha.m. On full harnessing and mobilisation of these water resources, say, by 2025 AD, (techno-economically feasible development) it is envisaged that 70 M.ha.m. of surface water and 35 M.ha.m of ground water can be mobilised for consumptive use. The projected use of this 105 M.ha.m of water is 77 M.ha.m for irrigation and 28 M.ha.m for domestic and industrial water supply and all other purposes. The envisaged water-balance on full development is depicted in a simplified form in Table 10.1.

It is relevant to emphasise that the return flow from ground water (45 M.ha.m) is the main contributor to the dry-weather flow in streams and rivers. It is also important to draw attention to the postulation that in the ultimate stage, 25 M.ha.m of ground water is to be contributed by surface irrigation. It is on this premise that 35 M.ha.m of ground water has been considered extractable annually for consumptive use. If in the long run, all the canal system is progressively lined and more efficient methods of irrigation like sprinkler and drip irrigation are adopted, the contribution from irrigation to ground water will be substantially less and the annually extractable ground water may be only of the order of 20 or 25 M.ha.m. Any over-extraction of ground water will correspondingly reduce the precious dry weather flows in streams and rivers and will

TABLE -I

WATER BALANCE ON FULL EXPLOITATION (SAY. 2025 AD)

[All figures in Million Hectare metres (M. ha.m.)]

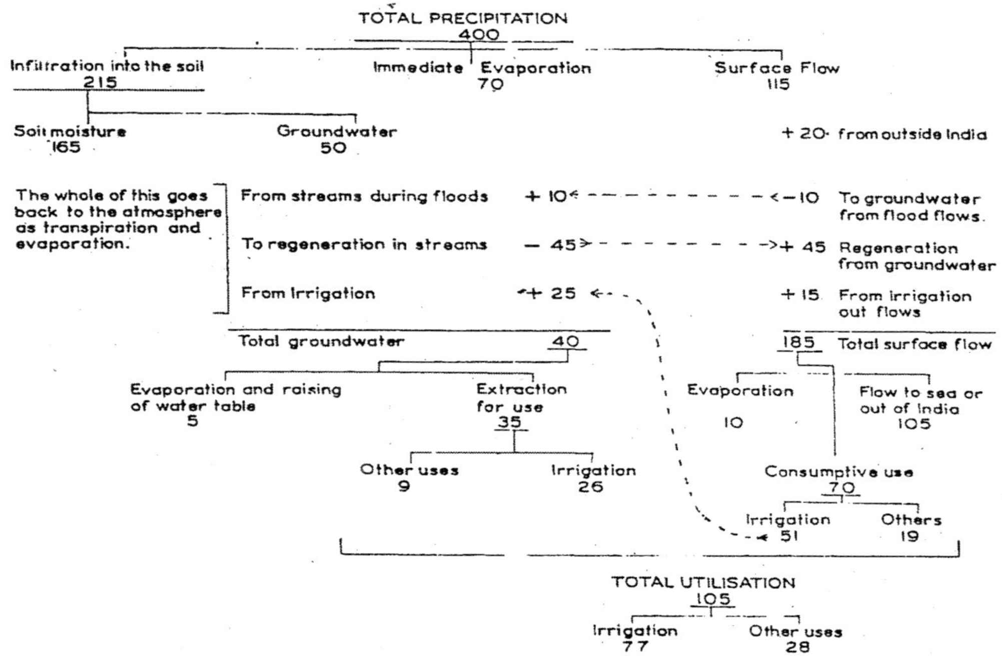

also lower the permanent ground water table, thereby upsetting the ground water regime. The often propounded idea that there exists an infinite quantum of ground water which can readily be extracted for consumptive use without any detriment to the surface water regime does not have any scientific backing. Another such theory is that by pre-monsoon overdrawal of ground water from the Gangetic plain, the whole basin can be used as a natural under-ground storage reservoir. However, the fact is that the rate at which surface water infiltrates into the ground is governed by the permeability of the top soil and the driving head (depth of water standing on the ground). As long as these parameters are not altered, the ingress of water into the ground cannot be increased by any extra lowering of the ground water level by pre-monsoon overdrawal.

Consequent to the seasonal concentration of rainfall, the river flows are also concentrated (about 85 per cent) in the period June to November, and that too during four or five flood spells of 5 to 10 day each. On the other hand, during the months of April, May and part of June, consequent to the progressively decreasing ground water return flow, the river flows dwindle to a trickle. For example the flow in Mahanadi at Hirakud (catchment area 83,000 sq. km. and mean annual rainfall 1370 mm) may remain more than 300 cumecs, occasionally peaking to 30,000 cumecs during July to October, but during the month of June, it dwindles down to 30 cumecs or less. This pattern is typical of all monsoon-fed rivers. Since the capacity of head-regulators and off-take canals or pipe lines are perforce very very small compared with the instantaneous flood-flows, the potential for direct diversion of run-of-the river flows for irrigation and other consumptive uses is very limited. The only tangible method by which a large portion of the average annual flows of the rivers can be mobilised for beneficial use instead of allowing it to flow down to the sea is to construct large reservoirs in the main stems of river, large enough to store the flood-flows and release it later in a regulated manner to meet the programmed uses. Since the major use is for irrigation, and the greater part of irrigation use is during rabi and the dry weather, all of the irrigation water has to be initially stored. In the National Agriculture commission's postulation of ultimate utilisation of 70 M.ha.m of surface flow, the aggregate storage capacity requirement has been taken as 35 M.ha.m. In my judgment this is an under estimation and a larger storage (live storage capacity) of the order of 50 M.ha.m. will be required. If we recognise the fact all the storages so far created have an aggregate live capacity of only 15 M.ha.m and those under construction will add only another 7 M.ha.m (together 22 M.ha.m) it is difficult to visualise how and when the additional 13 to 28 M.ha.m of storage will be created. Again taking the specific example of Mahandi, the assumption is that 5 M.ha.m out of its average annual flow of 6.7 M.ha.m will be mobilised for beneficial use in its final stage of development (2025 AD). The already completed and under construction projects in Mahandi basin together add up to only about 1.3 M.ha.m. To create the additional required storage of 2 to 3 M.ha.m, we need to construct four or five reservoirs of the size of Hirakud (live capacity of Hirakud is 0.584 M.ha.m) which perforce will involve large submergence and displacement of people. Many other large reservoirs are proposed on the Himalayan rivers, where the displacement of people will be less but has other adverse factors like high seismicity, less competent geological formations and forest submergence. If the on-going public objection to Tehri dam is sustained despite its technical soundness, no other storage project in the Himalayan foot hills is likely to come through. If the 226 m high Bhakra dam could be constructed in equally seismic and friable rock formations and is seen to be safely functioning for over 30 years, there should be no difficulty in arriving at a safe design for Tehri dam. But the environmentalists, sociologists and some public men are objecting to all reservoir projects on one ground or another whether the project be in the Himalayas, Central India or in Kerala in the remote south. The main points raised by them are :

(1) Large reservoir project cause large submersions and destruction of forest.
(2) They displace large number of people who are not adequately compensated or properly resettled.

The environmentalists argue that natural forests, particularly ever-green forests, are repositories of millions of years of floral and faunal heritage and posterity will be impoverished, if not driven to extinction, by the destruction of this genetic heritage. The shooting down of the Silent Valley hydro-electric project in Kerala is now world famous. The factual position is that the Silent Valley dam and the project colony, as proposed, were entirely outside the forest area. Its reservoir was to submerge 830 ha of forest land at the southern edge of the 9000 ha Silent Valley forest or, more precisely, 39000 ha of three contiguous forests.

The then Kerala Assembly in 1978 had unanimously resolved and assured the Government of India that no encroachment into the forest area other than the reservoir submergence will be done, that the entire forest complex shall be declared as a National Reserve, and that its preservation shall be ensured by a public-represented committee. Based on these facts and assurances, the then Government of India (Prime Minister Mr. Morarji Desai) cleared the project for construction. It is unimaginable how anybody can argue that the submergence of 830 ha at one edge of 39000 ha of contiguous forest area can destroy all the floral gene pool and also cause the extinction of lion-tailed macaques. Yet the so-called environmentalists and scientists obviously mis-presented facts and got the next Government (Prime Minister Mrs. Indira Gandhi) to veto the Silent Valley Project. The environmentalists have since been gloating that they have won the first major battle and that they will fight and win against other projects also. It has never been explained how the stopping of reservoir projects will improve the environment or living standard of a large part of the Indian population (some 300 million) who are below the poverty line and are living in filth and smoke.

They also argue that destruction of forests drastically reduces the incidence of rainfall and increases the floods and devastation, ultimately leading to desertification of the whole country. The scientific fact is that forests have hardly any influence on monsoon rainfall or devastating floods. Forests have grown in areas of plentiful rainfall with no human interference or other destructive forces. Forests certainly did not precede rainfall. Devastating floods occur when antecedent rainfall has already saturated the ground and then a very heavy rainfall follows, the whole of which runs off into the streams and rivers without any further detention. Some of the environmentalists were categoric that the low rainfalls in upper India in 1986 and 1987 was the direct result of deforestation in the Himalayas. However, when more than normal rainfall was received in 1988 and 1989, no explanations were forthcoming. Another misconception is that forests contribute to a larger ground water recharge. The fact is that with its deeper root zone, the infiltrating water is sucked back even at depths of 8 to 10 m, drastically reducing the quantum that goes down to the ground water.

It is a recognised fact that during the last 50 years, we have lost about 50 M.ha of forest cover, and that the loss is continuing at the rate of 1 to 1.5 M.ha per year. Forest land so far utilised by all the irrigation and multipurpose reservoir projects is less than 2 M.ha and that required for future projects may not exceed another 2 M.ha. It is therefore un-understandable and irrational that reservoir projects are being obstructed instead of tackling the immensely larger annual loss of 1 to 1.5 M.ha.

A facile and misconceived idea vehemently propagated by environmentalists and some administrators is that instead of large reservoirs, a number of small reservoirs and rural ponds can provide the same benefits without detriment to the forests and environment and without displacing people. It can be readily seen that if large reservoirs having depths of 50 to 100 m are to be replaced by small reservoirs with mean depths of, say, 10 m and rural ponds with mean depths of 3 m, the land area going under submersion will be 10 to 15 times larger. Where can we find such large areas of barren land (without forest and without people) which are suitably located to command the identified irrigation areas and concurrently capable of providing hydro power. flood control and other such benefits.

A softer and more conciliatory recommendation is that in response to protests, small reduction in the heights of dams (say, 10 per cent reduction in the depth of the reservoir) may democratically be accepted and thereby achieve an appreciable reduction in reservoir-submergence. This presumably supposes that dam-heights are more or less arbitrarily fixed and that the reduced height can still provide all or most of the envisaged benefits. This is, of course, not true. The dam height is actually fixed on the basis of careful analysis of incremental costs and benefits and submergence involved. In the case of the three large dams which are presently being protested against, the percentage reductions in submergence areas and live storage capacities (utilisable storage capacities) due to 10 per cent reduction in the depths of those reservoirs are tabulated below:

	Sardar Sarovar	*Indira Sagar*	*Tehri*
per cent reduction in submergence area	42.7	43.6	10.7
per cent reduction in live storage volume	57.2	61.6	30.0

These large reductions in the live storage volumes will correspondingly reduce the irrigation benefits and more than proportionately reduce the power generation because of the concurrent reductions in the operating heads of the respective power houses. On the other hand, the overall cost of the projects will only reduce by 12 per cent to 15 per cent and these projects will become unviable with the reduced heights.

As regards displacement of people, it is true that in the past the disbursal of compensations and rehabilitation of oustees were far from satisfactory. However during the last ten years, the provisions made for compensation and rehabilitation have been substantially improved. The question that still remains is whether these will actually be implemented and in a timely and compassionate manner. This is a matter which falls under the purview of the revenue department and the state government. However, it might be a good idea to constitute non-official committees with public representation to monitor and ensure an equitable implementation of the rehabilitation measures. The people who are to be displaced should feel assured that their interests will be safeguarded so that they will willingly move and not be forcibly evicted.

Despite the fact that large reservoir projects are obligatory for the development of our water resources, the prevailing mood at present seems to postpone or give up their construction. Under these circumstances, the National Water Plan formulated by the erstwhile Ministry of Irrigation to interlink the major rivers of the country and transfer water from surplus to deficit basins appears to be a non-starter. This proposal has the added constraint that under the Indian constitution, water is a state subject, and the enabling provision under Entry 56 of the Union List by which the Central Government could have taken over the regulation and control of some or all of Inter-state rivers and river valleys has not so far been invoked. As time passes, the chances for such a legislation being passed by the Parliament is becoming more and more remote. It is unrealistic to expect any State Government to voluntarily agree to transfer water from its territory to another state without a statutory pressure from the Union Government.

Does this mean that the future is totally bleak? I should think that as far as the agricultural sector is concerned, the future is quite bright. The United States of America, despite its gross land area of 937 M.ha and annual river flows approximately same as that of India, irrigates only about 20 M.ha; yet their annual foodgrain production is over 250 Million tons. We have already created an irrigation potential of 70 M.ha and the projects under construction will add another 15 to 20 M.ha. The lacuna is that our mean annual yield of foodgrains from irrigated cultivation is still less than 2 tons per hectare. With even partial improvements in our irrigation system and with some measure of incentive and financial assistance to the farmer to ensure adequate inputs of fertiliser and pesticides, 90 M.ha of irrigated area along with 50 M.ha of rainfed cultivation can easily produce enough foodgrains, pulses, oilseeds and everything else that are required to feed a population of, say, 1200 million in 2025 A.D.

The consumption of water for domestic and industrial water supply and other purposes (other than irrigation) was only 3 M.ha.m in 1974. Presently it may be about 6 M.ha.m. The projection that this requirement will increase to 28 M.ha.m in 2025 is an over-estimation. About 12 to 15 M.ha.m would certainly be sufficient, and there will be no difficulty in mobilising this much water. The constraint will actually be on the financial outlay required to pump the water over long distances to reach the people dispersed mostly in rural areas.

Holding up of large reservoir projects will, however, produce a serious constraint on hydro power development. Hydro power so far developed inclusive of projects under construction is only about 17 M.Kw out of the full estimated potential of 84 M.Kw. Tangible additions can only be made by large reservoir projects and not by run-of-river or micro hydel projects. Installing more and more coal-fired and gas based thermal power stations is techno-economically and environmentally far less favourable. Are we wise in rapidly exhausting the non-renewable resources of coal and gas and in the process severely exacerbating the atmospheric pollution? Will we at a later stage reach a dead-end and will have to revert to developing our hydro potential? With our population being already some 800 million and with about 40 per cent of them being landless and living below poverty line, how do we hope to prevent encroachments and destruction of forests? How do we intend to provide food and amenities to increasing millions of our people without irrigation and power generation? I am sure that posterity in whose benefit the reservoir projects are being curtailed or stopped will blame us for not having developed our water resources when it was easier and more economical to do so.

11

Large Dams and Cost-Benefit Analysis : Some Analytical Aspects and Policy Implications*

Kanchan Chopra

I INTRODUCTION

Resources in general and water resources in particular are once again occupying the attention of planners all over the country. This is, to some extent, inevitable in a country where agriculture is still the source of livelihood of a large percentage of the population and it is accepted (with some qualification) that irrigation has been the single factor responsible for increasing agricultural yields. Further, water availability in this country has now come up against a supply constraint. With the magnitude and distribution of water being what it is, exploitable water in this country is at best, limited. Whereas the vast arid and semi-arid regions in the western, central and southern parts suffer from an overall shortage of water, problems of waterlogging and poor drainage plague the eastern part. Moreover, high costs of inter-basin transfer make the option of transfer an infeasible one.

On balance, a number of problems plague the water resources development scenario. The focus of the discussion on water resources in this paper is, however, not any of these. The point at issue is the environmental impact of the technological options available for exploitation of water. This perspective makes the issues being discussed more complicated. The consequence is that contestants in the debate often talk at cross-purposes. One group 'believes' in small dams, the other 'believes' equally emphatically that as a technical option, these are

* The author is thankful to Professor Gopal K. Kadekodi for comments on an earlier draft and to Professor B.D. Dhawan for helpful discussions.

inefficient. Not surprisingly, the latter are those people, who, by and large, have had some concern with the shaping of irrigation policy in the country whereas the first group is of the environmentalists. Both groups have come to view cost-benefit techniques as an important decision criteria in terms of which dam projects should be evaluated. Consequently, internal rates of return and benefit-cost ratios have, of late occupied pride of place in discussions on environmental effects of projects. The technique has found favour with policy makers, primarily because it throws up a single number which can be used as the basis of decision making. It is, however, extremely sensitive to informational and methodological inputs which may include parameters that vary widely. Further, there exist at least three areas in which conceptual problems exist which cost-benefit analysis deals with insufficiently. These are:-- evaluation of environmental resources, the issue of inter-generational equity and sustainable development and the larger question of the nature of the development process and the closely related inter-personal equity question. The following sections shall attempt, among other things to review the state of the knowledge in these areas and deduce policy prescriptions in the particular context of large dams.

II INFORMATIONAL AND METHODOLOGICAL REQUIREMENTS OF PROJECT EVALUATION : COMMENTS BASED ON A CASE STUDY

Every project evaluation exercise is as useful as the informational input that goes into it and as accurate as the methodology it follows. This is best illustrated by an examination of a project evaluation exercise. In the context of large dams, Paranjapye's recent study of the Tehri dam, incorporating both a criticism of the exercise conducted by the Tehri Dam Project Authority (TDPA) and giving his own analysis provides a good starting point. His criticism of the TDPA's evaluation can be summarised as follows :

1. The rainfall and run-off figures are inaccurate as the snowbound part of the catchment has not been accounted for and the period over which the data has been taken is very short.
2. Accurate estimates of the rate of siltation, which have not been made, would decrease the life of the project from 100 to 62 years.

---

1. For the detailed study see Paranjapye (1988).
2. See Government of Uttar Pradesh, (1969), (1983), (1986) for the project reports.

3. There is a high probability of an earthquake of an intensity higher than provided for by the TDPA decreasing the life of the dam further unless a change in design, which increases the cost by 25 per cent is undertaken.
4. The assumption of a peaking power availability of 87 per cent is made whereas a Planning Commission review of the power situation has surmised that the peaking availability of hydel power stations is about 70 per cent.
5. In addition to these informational lapses, methodological ambiguities also exist in the TDPA evaluation. There is some arbitrariness in the distribution of costs between irrigation and power. The distribution is altered, it is felt, to show the B/C ratio for irrigation in a better light. The environmental benefits said to occrue from the catchment area development plan are also grossly overstated. The methodology used for finding the benefits from irrigation is questionable as loss of output, both on land submerged and land acquired for rehabilitation of oustees is not taken into account. The high yields assumed to hold on irrigated land are questionable and do not take account of the natural rate of growth of yield on this land in the absence of this project. Separate rates of interest for the irrigation and power sectors are taken which is not defensible. A uniform rate should be taken which should equal the opportunity cost of capital to the economy. The B/C ratio, arrived at by Paranjapye, after correcting for the above-mentioned lapses in the TDPA evaluation is 0.56.

The project analyst's predilections, however, tend to enter into all evaluation exercise. In the exercise being considered presently, for instance, Paranjapye is ambiguous about the power component of the project on distributional grounds. He argues "it would be a fair approximation to say that the benefits to society from additional power can be represented by the 'monopoly return' the government is extracting from consumers of power". It is however, argued in the literature that, in India, the government is subsidising power and providing it at rates which are lower than the resource cost of production. What is the correct position on this?

Three sets of prices exist and are significant in decision making in most economies. These are market prices, administered prices and accounting (or social) prices. Market prices are those that come about as a consequence of the operation of the forces of demand and supply. When the government fixes prices (as in the cases of power or irrigation water) we have a set called administered prices. Accounting prices are those that reflect the resource cost of producing an input or output when markets are imperfect. Administered prices, if they are lower than accounting prices, (as in irrigation water and power) involve an element of subsidy. In the absence of a market price, a monopoly return can only be exacted if the administered price is higher than the resource cost of production. One approximation of resource cost is the cost at which a commodity can be produced by the next best technological option or imported. An estimate of the approximate resource cost of producing thermal power (the alternative of hydel) was made by us. Using data on the inputs that go into the production of power[3] (coal, oil, operation and maintenance, interest on capital, administration and establishment) and using the resource cost of coal and oil we get a price of 82 per kWh. of electricity generated as the accounting price of power in Uttar Pradesh in the year 1987-88. This price does not take account of transmission and distribution costs to be comparable with Paranjapye's and the TDPA figures. As the TDPA's administered price has been 48 p. per Kwh, there is indeed an element of subsidy involved in the price at which power is sold.

The above line of argument assumes, of course, that power is a 'desired' commodity in accordance with the dictates of the current pattern of demand in the economy. It could be maintained that since it is, in general, consumed by a section of the people who enjoy higher incomes, its production is not as 'desirable' as that of irrigation water which is consumed by people with relatively lower incomes. This is however, a distributional argument and the literature on distribution weights should be used to determine the relative weight to be given to the consumption of different groups of people.[4] Further, power is not entirely divorced from rural economic activities. The districts scheduled to get irrigation water from the project in question have a network of

3. The estimate is made using the resource cost of all inputs that enter into the production of thermal power. The input coefficients and cost data are taken from Government of India (1988).

4. The exact distribution weights would depend on the income of the group relative to the national average, the leasticity of private marginal utility of income and the inequality aversion parameter of the government. See among others, Squire and Vander tak (1975) and Deepak Lal (1972).

tubewells. Shortages of power would affect their working. The efficient working of tubewells also decreases the possibilities of water-logging and hence the likely expenditure on drainage schemes, a necessary component of well-designed canal irrigation. Such conjunctive use of surface and ground-water illustrates the significance of the links between power availability and water-resource development. As power shortage are one of the major causes of underutilisation of tubewells in Uttar Pradesh, these interlinkage acquire a practical significance.

To take account of the points made above, we did our own calculations using UNIDO methodology and the social prices obtained by giving distribution weights to different commodities and to different categories of consumers.[5] The comprehensive B/C ratio based on an evaluation of irrigation benefits using the 'with' and 'without' principle is 0.87 (the IRR being 10.38).[6] The project continues to be not viable even after the UNIDO methodology has been followed.

While the informational basis of B/C analysis should underiably be accurate, the methodology should view the project in the context of its linkages with the rest of the economy. Figure 11.1 (see page 190) illustrates some of these linkages in the case of large dam projects. Effects of large dams on the economy are traced through their effects on land-use patterns, consumption of irrigation water and of power to effects on different sectors of the economy. It is clear that any comprehensive analysis of a large-dam project should examine each of these components and then arrive at an overview of the impact on different sectors and on the economy as a whole.

III EVALUATION OF ENVIRONMENTAL RESOURCES

When the stream of benefits and costs of a project spreads beyond the immediate context of the project, we often say that the effects are of

5. The weights used are :-- Power is given a weight of 1.2 as it is in short supply in the economy. Further, consumers of power are given a weight of 0.8 and of irrigation water a weight of 1.2. These are in line with weights used for other project evaluation exeercises and are derived as explained in studies referred to in footnote 4 above.
6. The above figures must be qualified by saying that they are derived from data as obtained from Paranjapye's monograph and not from the original project reports. The excercise was essentially an illustrative one.

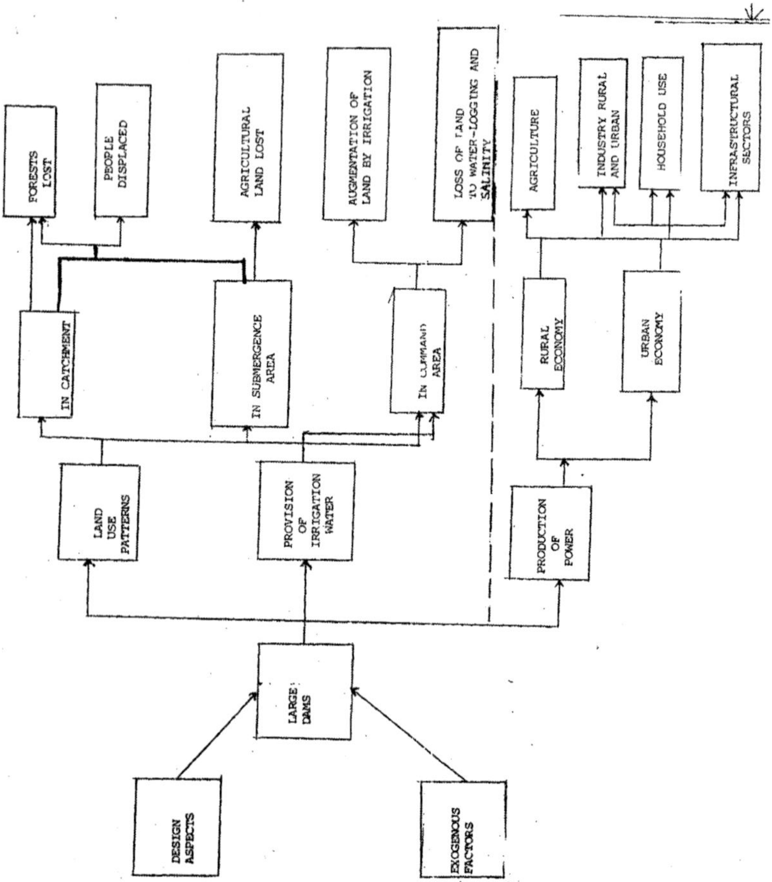

Fig. 11.1 : Large Dams : Effects and Linkage with other Sectors.

the nature of externalities. Most environmental effects fall in this category e.g. effects on forest area, groundwater levels, water-logging, or salinity.

A fundamental question that arises in the context of such externalities is : Is the benefit or cost from all addition or depletion to environmental resources the same? Alternatively, in the language of project evaluation, is the shadow price[7] of a unit of enviromental resources always constant. It is accepted now that this shadow price is not correctly reflected in the market price. Further however, does not their shadow price or value to society depend on the amount of their stock in the economy relative to need for it?

It shall be argued here that the value of an additional hectare of land under forests or the cost inflicted on society by a certain decrease in ground-water levels, (to give examples of two environmental resources crucial in large dam project) depends, as in the case of other resources, on (i) the stock of the total resource in the economy (ii) the value of land in alternative uses. Both these factors together determine the capability of the economy to assimilate the so-called 'environmental shock' that it suffers.[8] If the negative environmental effects of a project are within the range determined by this assimilative capability, the process of natural regeneration shall ensure that the stock reaches its initial level again. Use of conventional project selection criteria may, however, not ensure that the level of output opted for corresponds to this range.[9] Further, beyond a certain point on the 'environmental shock function', irreversibilities operate; i.e. the shadow price of environmental resources increases indefinitely.

Let us see what these arguments would imply in the specific context of deforestation caused by large dams. In view of the alarming pace of deforestation and the increasing need for forest resources for

---

7. For details on the methodology of project evaluation and definitions of concepts, see Dasgupta and Pearce (1972) and Dasgupta et.al (1972).
8. This concept is analogous to the so-called waste assimilative capacity of the environment which has been referred to in the literature for sometime now. See, for instance Pearce (1976) and Pearce (1988).
9. Pearce (1988) shows that the ecologically sustainable output for renewable resources can be assured even when project evaluation techniques are being used if the objective function is that of maximizing net social surplus over infinite time with rates of discount less than or equal to resource regeneration rates.

different uses, it can be argued that the stock of forest resources is less than optimal and that the level of their exploitation would certainly be higher than the level warranted by natural regeneration. The protagonists of forest preservation are, therefore, on firm analytical ground when they raise an alarm about the increasing pace of deforestation.

One has, however, to consider the fact that a hectare of forest land has a different value in different agro-climatic regions of the country. Forests are, in the final analysis, a kind of land-use and the value attached to them would depend on the correct land-use pattern for a particular region. In river valleys, for instance, cropped area may be a very large component of total land area and the proportion of forest area may be small and this may comprise the optimal land-use pattern. In mountainous and sub-mountainous regions, on the other hand, it may be essential, in the larger interests of ecological balance, to allocate a large proportion of land for forests. Similarly, grasslands may comprise the optimal land-use in areas of low rainfall which are best suited to livestock rearing from the agro-climatic viewpoint. Optimal land-use has to be defined in relation to the agro-climatic characteristics of each region in order to result in a balanced total use of land in the subcontinent. For each large dam, it has to be seen what the opportunity cost of the hectare lost to forests is. In general, the point at issue is whether the environmental loss (a) is within the capability of natural regenerating system to make up. (b) Comprises a rising social cost not taken into account by market forces. (c) is completely irreparable on account of the existence of irreversibilities.

The social cost of environment resources as a function of their use in the economy can be approximated by the curve OCC' in Fig. 11.2.

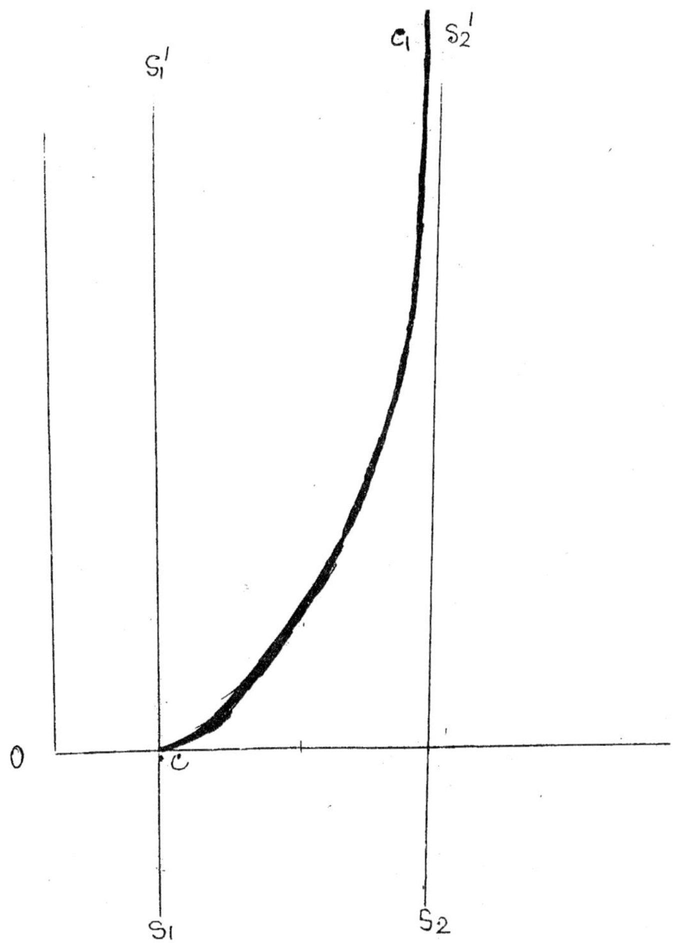

Fig. 11.2: Social Cost of Environmental Resources.

The line S_1 S'_1 stands for the regenerative capacity of the environment. Any use of environmental resources up to the point C, therefore has a zero social or resource cost. After C, the social cost increases and as use reaches the level indicated by the line S_2 S'_2, the region where irreversible damage is being inflicted on the environment is reached and the resource cost rises to infinity.

IV SUSTAINABLE DEVELOPMENT : INTERGENERATIONAL EQUITY

Another view on environmental resource management maintains that they should be kept intact for future use in order to ensure sustainable development. Two parts of this proposition can be subjected to further investigation. First of all, what is sustained development? There has, of late developed a considerable volume of literature on this. The World Commission on Environment and Development (WCED), for instance, defined substainable development somewhat simplistically as the capability to maintain present levels of output without affecting future levels.[10] Conway,[11] in a more clarificatory manner, maintains that sustainability is resilience within agro-ecosystems to stand not only stresses but also shocks such as drought and pests. It must be noted that such a definition does not imply that 'the development vector' increase monotonically over time. It only means that the trend must be positive.[12]

The second part of the proposition is that the stock of environmental resources must remain constant in order to ensure that trend increase in the development vector which is conceived of in the above definition as crucial to sustainability. This implies the assumption that such resources have a constant productivity and a constant demand, and that there is no technological change which makes them (the forests and the soil for instance) more productive or less demanded than they were. To elaborate, technological change or market conditions do not affect their productivity relative to that of substitutes, leaving relative demand unchanged.

With the assumptions mentioned above kept in mind, there are ways in which the future stock of environmental resources can be given a higher weight than is done in the project evaluation paradigm of investment choice which implies discounting the future. The first is by using lower social rates of discount so that the future gets a higher weight as compared to the present. However, a lower social discount rate, implying lower opportunity costs, will lead to higher levels of in-

10. See World Commission on Environment and Development (1977). This is, in a way, begging the question as the crucial issue is : what 'levels' of output?
11. Conway (1986) outlines this difference between shocks and stresses very clearly.
12. This is also the definition of 'weak sustainability' given by Pearce et. al. (1988). Note that development is conceived of as a vector with a number of attributes.

vestment and hence more use of inputs, among them environmental resources. This consequence illustrates the fallacy of applying at the macro-level a criteria which seems acceptable at the micro level.

Alternatively, in the overall context of sector planning, the compensating project approach could be adopted. If a particular project impinges on forest or soil resources, it could be compensated for by undertaking, simultaneously, another, the effects of which are in the reverse direction. Doing this would help to keep the total stock of environmental resources constant. It is said, for example, that more land could be planted to forests if a large dam project implies that forests have been cleared.

The process of use of enviromental resources may, however, either be irreversible or be difficult to compensate for due to several reasons. An input that is required both for the original project and the compensating one may be in short supply. Land, for instance, may be required both for the environmental resource depleting project and its complement, the resource creating one.

More importantly, every investment project (to be successful) envisages both technological and institutional frameworks of reference as inputs. Non-market structures may have provided the institutional superstructure before the project using environmental resources was introduced. They tend to be replaced by more commercially oriented institutions. Large dam construction projects in the hills, for instance, result in some deforestation. Simultaneously, they also erode the community based institutional structures that help to maintain them. When we argue for compensatory afforestation, we must keep this in mind and ask the question : Does this afforestation depend crucially on people's involvement? The latter may require a community orientation, a requirement which is inconsistent with growing commercialisation and market orientation, a consequence of earlier policy decisions. The correct policy intervention in this situation would aim at ensuring that institutional structures appropriate to the compensatory projects be sustained.

V QUESTIONS OF EQUITY AND THE NATURE OF DEVELOPMENT

The equity question is another major concern of the critics of large irrigation project; it is stated that the benefits of such projects accrue to one group of people while the costs are borne by another. Some commentators have gone to the extent of saying that the environmental question is actually a matter of the nature of the development process.[13] Two issues must be considered in this context. First, is the possibility of incorporating distributional issues in the methodology of project selection. Secondly, is the existence of the political will to do.

With regard to the first issue, project analysts have for long advocated the use of distributional weights in arriving at the total net benefits from a certain project. This would imply taking a weighted average of net benefits, the weights being inversely proportional to the level of income of the beneficiaries. Detailed procedures on the methodologies to be followed to determine these weights exist.[14] In essence, they imply taking either society's preferences (as expressed in individual behaviour patterns) or government's preferences (considering it a custodian of society and representing its preferences through its tax-subsidy structure), as the set of values to be accepted in determining them. The assumption implicit in using the latter is that the government is a representative of society. However, there exists, now a large volume of literature on how it is inappropriate to do so. The image of the benevolent government has been eroded substantially. In its place have emerged the paradigms of government as a Leviathnic monstrosity and as a rent-seeking entity.[15] To support these analytical perspectives, there exists a growing body of studies that codifies, both in qualitative and quantitative terms, the exact role of governments as rent-seeking entities and its implications for resources use and allocation.[16] It follows, therefore, that income distribution weights should be derived from individual behaviour patterns rather than government preferences.

In the field of income distribution, it has to be understood, however, that value judgments are imperative when it comes to policy making. The analyst can only make explicit the impact of a certain project on the income levels of different groups of beneficiaries.

13. Recent statements of this view are found in papers in this volume. See among others Iyer (1989).
14. For a description of the methodology of determining distribution weights see, Squire and Vander tak (1975), Deepak Lal (1972) and Murti (1980)
15. See Buchanan (1975).
16. For an excellent account of such rent-seeking behaviour in the context of canal irrigation in South India, see Wade (1982).

Another analytical tool that can help in this is the "income impact matrix". This matrix gives in detail, the consequences of a project for different income groups. A recent case study adds an 'environmental damages' vector to this matrix and builds up a community impact index.[17] This index which measures the preferences of affected residents may result in a policy decision which reverses that arrived at by a project evaluation exercise. All that such a result would serve to elucidate are the normative value judgments underlying different ways of viewing the same project. This is precisely what an analyst's thrust should be. It is then upto the political process to take up postures based on value judgements and to authenticate or nullify such stands through popular expression.

VI POLICY SUGGESTIONS

The main thrust of the suggestions for policy-makers emerging from the above analysis is in three directions. First of all, of course, there is need for more informed benefit-cost analysis. The informational input that goes into project evaluation exercises, must be more accurate, comprehensive and complete. With all their shortcomings, project evaluation exercises still have a positive contribution to make in concretising criteria for investment allocation. Such project evaluation exercises should not hesitate to make explicit the trade-off or, alternatively, the complementarity, between environment and development that is implicit in them and then state which part of the environmental degradation function the particular project being studied lies on. This can both help policy-makers and educate public opinion. Next, water resources planning in the country should be based on an agro-climatic regional approach. The social value of switching land from one use to the other should then be determined separately for each region. Following from this, the value of a hectare of land lost to forests due to the setting up of large irrigation projects or to cultivation on account of water-logging and salinity consequent on spread of canal irrigation can be determined. The total impact of a set of projects

17. See Rose, Stevens and Davis (1989). The empirical results for a project studied are revealing. The community impact index reveals that a majority of residents of the MNF (the project in question) may oppose the mining policy despite the fact that the gross gains are three times the gross losses. "This result emerges out of the fact that environmental damages are rather evely distributed while the gains are skewed in favour of the upper income residents." For projects where environmental damges are regressive in impact, the community impact index would have such a consequence.

in a particular agro-climatic region on its environment can then be worked out. In the context of water resources, it would be useful to distinguish between the following three agro-climatic regions.

1. The high rainfall regions of the east where drainage and water-management problems are the most pressing ones. The excess of water consequent on poor drainage results in a low production impact of irrigation. Though groundwater exploitation is one solution to this problem, it implies that some institutional changes be made. In terms of technological options, capturing water flowing into the already water abundant eastern region in the upper catchments themselves has been indicated.[18] It is clear that both large and small reservoirs shall have to be commissioned to do this. Information intensive project evaluation which makes sure that the negative environmental impact is clearly identified should be made a must for all such projects. Compensatory afforestation projects, where undertaken, should have in-built mechanisms for the creation and nurturing of institutional frameworks that help in making them a success.
2. The low to medium rainfall plateau regions of central, western and southern India where tremendous scope exists for small-scale watershed management and soil and moisture conservation projects.[19] These preserve that soil and harvest rain-water as it falls and result in increase in productivity of agriculture, and livestock rearing along with environmental preservation.[20] They comprise one large area where there does not seem to exist any conflict between efficiency, equity and environment preservation. All possible investments in this direction must be encouraged.
3. River-valleys where the water-irrigation intensive agriculture has in the last two decades resulted in large increases in production. It has been illustrated in some cases that, so far the impact of water-logging and salinity has been localised and limited and a balance between groundwater and surface water use has been attained.[21] Over-use of water, nitrate pollution and biomass imbalance could emerge as possible problems in such regions.

---

18. For a succint discussion of the problems of the eastern region see Chapter 12 of Dhawan (1989).
19. For a review of studies relating to such watershed management projects see Chopra (1990b).
20. See Rao (1989) for a similar viewpoint.
21. See, for instance, the study of Punjab agriculture contained in Chopra (1990a).

Careful monitoring needs to be done to view and correct for such effects.

In addition to information-intensive project evaluation and a comprehensive land-use policy based on agro-climatic regions, two kinds of empirical exercises need to be conducted for every project. Firstly, the institutional structure within which an investment shall take place and the changes it is likely to bring about in local mores and structures need to be spelt out. Secondly, an income-impact matrix including an 'environmental damages' vector should be constructed for the region in which the project is to be set up. This shall illustrate what impact the project is likely to have on regional incomes. Together, these two exercises shall enrich the informational content beyond that contained in B/C ratios or internal rates of return, which of course, are important in their own right. They shall clarify what impact the project has on people's livelihoods through developmental, institutional and environmental changes. Simultaneously the overview of land-use policy shall place environmental resources in their correct economic perspective. This informational background shall then give a further economic and social content to discussions on the environmental aspects of all project.

REFERENCES

Buchanan, J.M. (1975), *The Limits of Liberty : Between Anarchy and Leviathan,* University of Chicago Press, Chicago.

Chopra, K.Kanchan (1990a), *Agricultural Development in the Punjab; Issues in Resource Use and Sustainability.* Vikas Publishing House, New Delhi.

Chopra, K. Kanchan (1990b), "Watershed Management : Review of Recent Evaluation Studies", Mimeo, Institute of Economic Growth, Delhi.

Chopra, Kanchan (1987), "Social Forestry Projects : An Approach to Evaluation", *Resources Policy,* 13(2), pp. 142-158.

Conway, G.R. (1986), "The Properties of Agro-ecosystems" in *Agricultural System.*

Dasgupta, A. and Pearce, D.W. (1972). *Cost-Benefit Analysis,* Macmillan.

Dasgupta, P., Sen, A.K. and Marglin, S.A., (1972) *Guidelines for Project Evaluation,* UNIDO, Oxford University Press, New York.

Deepak Lal (1972) "On Estimating Income Distribution Weights for Project Analysis", *IBRD Economic Staff Working Paper* No. 130.

Dhawan B.D. (1989) *Studies in Irrigation and Water Management,* Commonwealth Publishers, New Delhi.

Government of India, (1988) *Annual Report on the Working of State Electricity Boards,* Planning Commission, New Delhi.

Government of Uttar Pradesh, (1969, 1983, 1986) *Tehri Dam Detailed Project Report* Vol. I, (Revised) and Vol. II.

Iyer , R.R. (1989) Large Dams : The Right Perspective, in this Volume.

Nijkamp, P. and Rouwendal, J., (1988), Intergenerational Discount Rates in Long-term Plan Evaluation, *Public Finance* pp. 195-212.

Pearce, D.W., Barbier, E.B. and Markandya, A. (1988) *Sustainable Development and Cost-Benefit Analysis", Paper for Workshop on Integrating Economic and Enviromental Assessment-Vancouver,* Canada, November.

Pearce, D. (1988) Optimal Prices for Sustainable Development" in Pearce, D., Collard and Ulph, O. (Edited), *Economics Growth and Sustainable Environments*; Macmillan.

Pearce, D. (1976), "The Limits of Cost-Benefit Analysis as a Guide to Environmental Policy" *Kyklos*, 29 (1), 97-122.

Paranjapye, V. (1988) *Evaluating the Tehri Dam : An Extended Cost-Benefit Appraisal,* INTACH, New Delhi.

Rao, C.H.H. (1989) "Technological Change in Indian Agriculture : Emerging Trends and Perspectives", Presidential Address. The Golden Jubilee Conference of the Indian Society of Agricultural Economics, *IJAE*, Vol. 44, No.4 Oct-December.

Rose, A., Stevens, B. and Davis, G., (1989) "Assessing Who Gains and Who Loses from Natural Resources Policy, Distributional Information and the Public Participation Process", *Resources Policy,* 15, pp. 2282-291.

Wade, R., (1982) "The System of Administrative and Political Corruption : Canal Irrigation in South India" *Journal of Development Studies,* 18/37, pp. 2287-328.

World Commission on Environment and Development, (1987) *Our Common Future,* Oxford University Press, Delhi.

12

Evaluation of Public Irrigation Systems

A. Sundar

1. Although, in India, irrigation has been practised for thousands of years, and, since Independence, a large number of public irrigation systems have been developed, there is no known case of any in-depth evaluation of the performance of public irrigation systems.
2. One reason could be that investments in irrigation systems, for use by the community, have been considered, all along in our history, as an act of piety, a philanthropic action, which needs no justification, and, so any evaluation. It was considered to be self-evident that the resulting benefits to the society were so significant that questioning it, through an evaluation, was akin to questioning the permanence of the 'soul'. So, we have engaged in irrigation for thousands of years on the basis of the self-evident beneficial nature of public irrigation systems.
3. But, now, the scenario has changed. There is considerable criticism from many sections of society that public irrigation systems are too expensive to be justified. They argue that it costs too much to the society and its benefits are not worth the cost. All these arguments also, however, are only based on impressions and subjective assessment, since no system has even been evaluated so far.
4. Could it be that the significant change, that the country has undergone in the past two to three decades of planned development, from an almost totally agrarian and rural society to a middle order industrial urbanised one, be responsible for this?
5. To one engaged in agriculture, in a monsoonic climate, asking to justify irrigation is like questioning the utility of ploughing fields. He considers any such enquiry to be a waste of time and

effort. But, to the person engaged in trade and commerce, industry, services and other mainly urban-based activities, public irrigation systems means a competition to the resources, which otherwise would have been invested in his surroundings. Agriculture to him is a source of his food (mainly), and, at present, the public distribution system makes sure of it, more, so since it is controlled by his class. Furthermore, with the reasonable assurance on the food front, what all he needs are the inputs for his activities, infrastructure for the same and, of course, the market.

6. Growing urbanisation and urban purchasing power assures a reasonable market, which can be served, if given the proper infrastructure and inputs. The main input being energy, irrigation systems have no role to apply. So, investments in irrigation, now, need full justification, more so, since investments in other sectors – developed basically on western thoughts, processes and procedures -- have a history of quantification of benefits and costs. When one wishes to set up an industry and tries to raise finances, he is asked to justify his project. So too in other urban activities. Why not, the same logic be applied to irrigation too, is a very legitimate question.
7. India, being influenced by the monsoons, has her rivers and streams flowing with high variability. Most of the streams are virtually dry during the summer except the snowfed rivers in the North and carry almost 90 per cent of their annual volumes in the period of 4-5 months, with the balance of 10 per cent carried in the remaining 7-8 months. Any diversion scheme can have only this amount of 10 per cent volume, spread over 7-8 months, so about 1.5 per cent of the annual flow every month, during the non-monsoon season, when Rabi has to be supported. Considering the country as a whole, this works out to about 3 million ha.m (1.5 per cent of 188 million ha.m), assuming the entire flow to be diverted. Since this would not be possible, and taking the diversion to be 50 per cent of available flow, all that can be diverted is about 1.5 million ha.m. This can sustain crops, even under the most efficient use, on only about 3 million ha. of land.
8. Some critics argue that we are allowing all the monsoon rain to run off to the oceans by not taking enough soil conservation and water conservation measures. They claim that, by proper soil and water

conservation measures, the soil profile can retain enough moisture to sustain a crop after the monsoon.

9. About 35 per cent of the country receives less than 750 mm of rainfall, on the average, while 35 per cent receives between 750 to 1150 mm of rainfall. Bulk of the rainfall is received, especially in areas receiving less than 1150 mm, covering about 2/3 of the country, in a few spells – about six or seven. During these heavy rainfall spells, the soil does not permit absorption of more than about 30 per cent or so of the rainfall to be used for crop sustenance. Therefore, any amount of soil conservation and other measures cannot enable more than about 30 per cent or so of the total rainfall to be retained in the soil during the rainfall season. The stored amount could be only less than 250 mm, in areas receiving less than 750 mm rainfall (35 per cent of the country), and between 250 and 350 mm, in areas receiving between 750 and 1150 mm of rainfall (another 35 per cent of the country). All that can be achieved by the retained moisture is a good kharif crop (if rain sets on time and is average or above it), and, in some areas, a little more of soil moisture carried forward to Rabi (of the order of about 150 mm). That amount certainly cannot sustain a good Rabi crop- even following a good monsoon. No doubt, however, soil conservation measures would ease problems of drinking water in many areas and minimise loss of top-soil, but, they cannot sustain agriculture in the post-monsoon period.

10. With the background, let us consider the need and the role of storage-based public systems of irrigation. Where rainfall is below 750 mm, i.e., in arid and semi-arid regions, no profitable agriculture is possible without irrigation, even in Kharif, let alone Rabi. In such areas, moreover, irrigation can be developed only by tapping rivers that receive their flows from heavier rainfall or snowmelt zone, as needed for irrigation in Punjab, Haryana and Rajasthan. Limitations on diversion being taken into account, storages are inevitable for such irrigation systems. Even in areas of heavy rainfall, if one wishes to overcome the inevitable uncertainties of onset of rains and breaks in it, storage of water is needed, since the low flows, before rains, are either virtually nil or very meagre. Storage is also needed if one wishes to raise Rabi crops. Thus, there is an inevitable need for a storage of water in view of the natural conditions in India, due to its geography. Such a need for agriculture is clear. But such storages are also needed

if one wishes to assure water for other uses, for example, industrial and domestic supplies which will increase in the future.

11. In enabling agriculture itself in arid and semi-arid regions, irrigation plays the role of 'support'. Without irrigation, these areas would not be put to profitable agricultural use. This 'support' role has to be performed continuously throughout the crop season. On the other hand, helping to overcome uncertainties is more of a role of insurance against, or 'rescue' from, a possible undesirable consequence, and is performed only when called upon to do so. This is the role of irrigation, in Kharif, in heavy rainfall zones.
12. Therefore, the first step in any performance evaluation would be to assess whether the assigned or required role is being performed. Once it is certain that it is so, the next step is to determine the outcome (agricultural production and productivity, as well as any undesirable consequences) of such a role performance.
13. In assessing each of the above two aspects, one has to constantly refer to the fact that the systems under consideration are public systems. Being public systems they have to be non-discriminatory, and all the farmers in the command should be treated alike. Should there be discrimination, it cannot be functioning as a public system.
14. What is to be assessed, hence, is whether the system has an operating policy that is derived from the objective of non-discrimination? Does this policy indeed work to achieve that goal? Comparison of quantities delivered and frequency of delivery to different groups would help quantify this assessment. If these variables lie within a very narrow permissible range, the system would be working as a public system.
15. Support role would have been performed adequately, if the system provided the promised number of planned irrigations, assessed from support requirements. Similarly, 'rescue' role would have been played fully if, at all the crises, the system provided 'rescue'.
16. The above assessment would tell us whether the public irrigation system has functioned according to its assigned role. But, this alone does not justify that the system should have been built or was wanted, which is what has been exercising many minds. So, now, we need the quantification of increased output from agriculture that has been made possible by irrigation. Assessing this, when

irrigation plays the rescue role might be more difficult than when it plays the support role, since virtually all the output in the latter case is enabled by irrigation. Offsetting this, mainly, are the losses due to submergence of land for storage, from the build up of ground water levels in irrigated areas and from the ill-effects of over-irrigation.

17. Ill-effects from over-irrigation can be virtually eliminated if the system can be operated well according to a well thought out policy and keeping low levels of intensity of irrigation. Certainly, the prevailing socio-political framework has a very significant role to play in this regard.
18. Increased recharge of ground water is inevitable, since the land under irrigation receives water (more of it) over a much longer duration than under natural rainfall conditions only. In fact, where water is transported from outside the area, it is equivalent to having increased rainfall manifold – as in Haryana, Punjab and Rajasthan. This could lead to beneficial as well as deleterious effects. Increased ground water levels permit more supplies (under one's control through wells) at a less expensive way (due to higher water levels) than before irrigation. Poor quality would be the problem, if the geology of the area is of that type – too much of undesirable salts.
19. But, the more currently publicised loss is from the effects of submergence. Submergence, however, is inevitable for storing water and the need for storages in our country has already been discussed. So, basically the question is how to minimise area submerged for a given storage, or rather for gaining a certain quantity of irrigation coverage – 'support' as well as 'rescue'.
20. There is a school of thought that propagates the idea of having a large number of shallow small storages (commonly known as tanks in South India), contending that, overall, this will be beneficial, not just in terms of area submerged alone, but also in other aspects, like dispersal over the country side.
21. In South India, tanks have been built and operated for several hundred years. They are well dispersed over the three South-Indian states of Karnataka, Tamil Nadu and Andhra Pradesh and number almost 120,000. But they have been unable to ensure the stability that large scale irrigation systems can and have provided. The reason is that both the catchment and command areas experience the same sort of rainfall conditions. When rainfall is

low or absent in the command, it is so in the catchment too. Thus a drought cannot be mitigated. Tanks can only help to carry forward a part of monsoon flow (whatever is) to non-monsoon months. Problems due to delays in monsoon cannot be overcome.

22. As far as land submergence is concerned, it is proportionately more in the case of tanks than in the case of major reservoirs, because tanks are shallow storages. In the case of tanks, compared to major reservoirs, on the average, the land submerged will be two to three times more, for the same irrigated area. The submerged area, in tanks, is about one-third to one-half of the command area, while in major storages this will be, utmost, of the order of 10-15 per cent.

23. An even more important consideration is that of carrying water from higher rainfall zones to arid and semi-arid regions. Certainly, tanks cannot be employed. Such movement, however, is absolutely needed in view of one-third of the country being in arid and semi-arid conditions. Here, only major storages can function.

24. Land incidentally, is taken out of agricultural use or forest use also by many other developmental activities – industries, transport, housing etc. In all these cases, land is only used as a substratum for various purposes, the potential of land to support vegetations being completely lost. However, irrigation system at least compensate to the extent, that productivity of the command area is multiplied, thereby virtually providing more land capability for vegetative growth than what was lost. One can say it creates more land! Even on a modest assumption of doubling of productivity in irrigated areas, in major scheme, with about 10 per cent submergence ratio (water spread area to command area), the additionality is almost 90 per cent.

25. However, it is true that there are environmental effects of losing some forested areas. But taken in the overall perspective, much more forest has been destroyed by industrial activities, particularly paper, pulp and other industries, including the insatiable demand for furniture. Considering at 10 per cent submergence ratio, all the major and medium projects put together have covered only about 3.5 million ha. of land (10 per cent of 35 million ha. of irrigation under major and medium schemes). Even taking about 75 per cent of the area submerged to be forested, the toll is only 2.6 million ha. Allowing for another 50 per cent despoiling due to influx of people during construction,

land occupied by the Canal conveyance and distribution system, the total is only about 4 million ha. It is everybody's knowledge that the forest cover lost in the period since Independence is 10-15 times the above figure. So making irrigation systems the villain-of-the-piece is not based on facts, but, perhaps, on some unexplained dislike to such developments, partly aided by reasonable self-sufficiency in food. This is also evident from the fact that deforested areas are mostly hill tops and slopes far removed from valleys which would be the sites of reservoirs. Industrial uses, one could say, have played havoc with our forests.

However, the question still remains. Are irrigation schemes justified? This can be answered only if careful assessment is carried out of at least one major project in each state.

26. Would the concerned irrigation departments initiate such studies? They need them if they wish to continue in business !